NUMERICAL APPROXIMATION

23-4-71

LIBRARY OF MATHEMATICS
edited by

WALTER LEDERMANN
D.Sc., Ph.D., F.R.S.Ed., Professor of
Mathematics, University of Sussex

NUMERICAL
APPROXIMATION

BY
B. R. MORTON

LONDON: Routledge & Kegan Paul Ltd
NEW YORK: Dover Publications Inc

*First published 1964
in Great Britain
by Routledge & Kegan Paul Ltd
Broadway House, 68–74 Carter Lane
London, E.C.4
and in the U.S.A. by
Dover Publications Inc.
180 Varick Street
New York, 10014*

*Second impression 1966
Third impression 1969*

© *R. B. Morton 1964*

*No part of this book may be reproduced
in any form without permission from
the publisher, except for the quotation
of brief passages in criticism*

SBN 7100 4354 6

*Printed in Great Britain by
Latimer Trend & Co Ltd, Plymouth*

Contents

Preface *page* vii

1. Introduction to Numerical Analysis — 1

    1.1. *The need for numerical methods of analysis* — 1
    1.2. *Approximations to functions* — 2
    1.3. *Numbers, continuous functions, tables of values and tabulated experimental results* — 7
    1.4. *Errors and accuracy in numerical analysis* — 9
    1.5. *Effective numerical computation* — 12
    Exercises — 13

2. Finite Differences — 15

    2.1. *Tables of values* — 15
    2.2. *Finite differences* — 17
    2.3. *Notation for finite differences* — 20
    2.4. *Finite differences of a polynomial* — 22
    2.5. *Detection of mistakes by differencing* — 25
    2.6. *Divided differences* — 28
    Exercises — 31

3. Interpolation — 32

    3.1. *Linear interpolation* — 32
    3.2. *Interpolation using divided differences: Newton's formula* — 34
    3.3. *Lagrange's interpolation formula* — 38
    3.4. *Interpolation in a table with equal intervals* — 38

CONTENTS

    3.5. *The use of Bessel and Everett interpolation; throwback*   44
    3.6. *Some uses of interpolation*   48
    3.7. *General notes on interpolation*   52
        Exercises   53

4. Differentiation and Integration (Quadrature)   56

    4.1. *Numerical differentiation*   58
    4.2. *The integration of a specified integrand; Simpson's rule*   62
    4.3. *Quadrature from a table of values*   67
    4.4. *Double and repeated integration*   75
    4.5. *Gaussian quadrature*   78
        Exercises   80

5. Method of Least Squares   82

    5.1. *Fitting straight lines to data*   83
    5.2. *Polynomial approximation by least squares*   85
    5.3. *Data smoothing*   90
        Exercises   92

Note on Operational Methods   93

Answers and Notes on Exercises   94

Directory of Numerical Processes   97

Index   99

Preface

NUMERICAL computation plays an important and growing role in the physical sciences, engineering and technology, and it serves across the whole range from the development of new theories to the analysis of experimental data. As problems increase in complexity and as high-speed computing machines spread so that their use is the rule rather than the exception, it becomes increasingly urgent that most students in these subjects should have some introduction to numerical techniques.

The effective use of computing machines involves two stages. First we need an understanding of the kinds of way in which numerical processes can be constructed and put to use; and then we must learn the system of programming for the particular machine at hand. Programming systems are like languages which enable us to describe problems in terms acceptable to the machine; at best they are remarkably easy to learn, but in the long run their effective use must depend on a proper preparation of problems and hence on a thorough understanding of numerical analysis.

In this short book (and a forthcoming companion volume *Numerical Solution of Equations*) I have chosen to concentrate my attention on the mathematical aspects, because my experience has shown that in the end students will master the use of fast machines more quickly if the early stages of teaching are separated, so that they are introduced first to the ideas underlying numerical processes, and then to their adaptation to machine use.

Like any other branch of mathematics, numerical analysis must be *understood* if it is to be of any real value.

PREFACE

For this reason I have discussed relatively few of the many formulae available (as long lists of formulae add nothing to understanding), and I have avoided the use of operational methods (as these often confuse inexperienced students by obscuring the significance of steps in the derivation of a formula).

The examples and exercises are particularly important, as it is difficult fully to grasp numerical processes in the abstract; they form a real part of the book, and the reader is recommended to follow all numerical steps using such calculating aids as may be available. Nothing is simpler than to construct fresh exercises from any set of tables.

B. R. MORTON

The University,
Manchester

CHAPTER ONE

Introduction to Numerical Analysis

1.1. The need for numerical methods of analysis

Progress in the physical, engineering and other sciences is based on two parallel but strongly coupled streams. On the one hand *experimental*, involving the design of experiments and the collection of physical data; and on the other hand *theoretical*, through the construction of model theories and their use in the interpretation of experimental results and in planning new experiments. The results of experiments are obtained as numbers or as tables of numerical values, and most theoretical solutions have also to be expressed numerically before they can be assessed critically by comparison with experiment.

The most important mathematical techniques used in the development of model theories are the normal processes of analysis, many of which are introduced in other books of this series. These are powerful processes which are of the greatest possible importance to the applied mathematician, but they often produce *formal* solutions rather than the final *numerical* solutions that are needed for immediate display of the physical significance of a solution and for direct experimental comparison. For example, it is easy to show that the solution to the differential equation

$$dy/dx + 2xy = 1 \text{ with } y = 0 \text{ at } x = 0 \text{ is } y = e^{-x^2} \int_0^x e^{t^2} dt,$$

but this formal solution is of little numerical value as it is easier to solve the original equation by a direct numerical process than to carry out a numerical integration of the rapidly growing integrand e^{t^2}. (The 'formal' solution *is* numerically useful given tables of e^{-x^2} and $\int_0^x e^{t^2}dt$, but these will have been calculated numerically and the use of such tables is itself a numerical process; moreover, this particular integral is quite likely to have been tabulated through a direct numerical solution of the differential equation!) There are other types of problem, such as the solution of non-linear differential equations, which are of central importance in current work on physical theories but which cannot yet be handled analytically to any general extent, and numerical methods of solution have an essential role to play here if theory is not to lag seriously behind experiment. Again, the normal methods of analysis cannot be applied directly to operations on tables of experimental values or other tables, and such operations must obviously be carried out using numerical processes.

Numerical work will often enter as a normal part of the mathematical treatment of a problem, and numerical methods form an essential part of the standard equipment of any applied mathematician.

1.2. Approximations to functions

A basic analytic method is the expansion of a 'complicated' function as a convergent infinite series of terms with simpler (or otherwise more appropriate) functional form. These expansions may be regarded as infinite approximation processes in which the error may be made arbitrarily small by taking progressively more terms into account.

The most important case is that of *Taylor expansion in an infinite power series*, where the function $f(x)$ is represented for some neighbourhood of a chosen point x_0 by the infinite series,

$$f(x) = f(x_0) + (x-x_0)f'(x_0) + \frac{1}{2!}(x-x_0)^2 f''(x_0) + \ldots$$
$$\ldots + \frac{1}{n!}(x-x_0)^n f^{(n)}(x_0) + \ldots$$

1.2 APPROXIMATIONS TO FUNCTIONS

in powers of the displacement $x - x_0$, where $f' \equiv df/dx$, ..., $f^{(n)} \equiv d^n f/dx^n$, and $f^{(k)}(x_0)$ is to be calculated at $x = x_0$. Such an expansion will be valid within some range of convergence $a(x_0) < x < b(x_0)$, and within this range the truncation error produced by discarding all terms after the nth can be made smaller in magnitude than any given positive constant by judicious choice of n. The difficulty from a *numerical* point of view is that this convergence may be so weak as to be useless in practice, and also it describes the ultimate behaviour of the terms in the series whereas numerically we want to concern ourselves with only the leading terms. Thus, the expansion

$$\tan^{-1} x = x - \frac{1}{3}x^3 + \frac{1}{5}x^5 - \frac{1}{7}x^7 + \ldots$$

is convergent for $-1 \leqslant x \leqslant 1$; numerically it is very effective for small values of x and only two terms are needed at $x = 0.1$ for five decimal accuracy, but the number of terms needed increases rapidly with x, to seven terms at $x = 0.5$ for five decimal accuracy and *fifty-one terms* at $x = 1$ for *two decimal* accuracy.

The Taylor series can also be written in the terminated form

$$f(x) = f(x_0) + (x - x_0)f'(x_0) + \ldots + \frac{1}{n!}(x - x_0)^n f^{(n)}(x_0) + R_n,$$

with *remainder* $R_n = \dfrac{1}{(n+1)!}(x - x_0)^{n+1} f^{(n+1)}(\xi)$ for some ξ in the interval (x_0, x). The remainder R_n may be regarded as the error involved in the approximate representation

$$f(x) \simeq f(x_0) + (x - x_0)f'(x_0) + \ldots + \frac{1}{n!}(x - x_0)^n f^{(n)}(x_0),$$

and it may be seen that this is a simple polynomial representation of degree n for the function $f(x)$. But can we reasonably expect that such polynomial representations can in practice be used without substantial error caused by neglect of the remainder R_n? This question is answered partially by Weierstrass's theorem on approximation by polynomials.

Weierstrass's theorem* states that *a function continuous in a closed interval can be approximated over the whole interval by a polynomial of suitable degree which differs from the function by less than any given positive quantity ϵ at every point of the interval*. As the allowable error of approximation ϵ is decreased, the degree of the approximating polynomial must be increased. This theorem is of the greatest importance because it implies that in practice, where we work always to a limited and specified accuracy only, we can replace any continuous function over a given range by an approximating polynomial without significant loss of accuracy. Hence we are justified in assuming a polynomial approximation for a given set of experimental or other values, and also in matching given functions by polynomial approximations over closed intervals. There remains one major drawback, in that *Weierstrass's theorem gives no indication whatever of the degree of polynomial required* for a given accuracy of approximation. This is typical of the difference in attitude in pure and in numerical analysis; the pure analyst is concerned first with the general principles of approximation, but the numerical analyst wants to know when he can reasonably use quadratic or cubic approximation! We shall see that in practice there are ways of judging the degree of a suitable polynomial representation even when no direct estimate for the remainder R_n can be made.

The way is now open for a major change in emphasis. We shall adopt unashamedly the idea of simple polynomial approximation to the functions or tabular values involved in calculations as a basis for developing numerical processes. Moreover, we shall reinterpret the significance of the re-

* For a proof and various extensions see e.g. **Jeffreys and Jeffreys** *Methods of Mathematical Physics*, Cambridge.

1.2 APPROXIMATIONS TO FUNCTIONS

mainder or error term of the terminated Taylor expansion as providing a guide to the working range over which a *polynomial representation* of specified degree can be used to the accuracy required. We shall thus construct numerical processes in which each function is represented by a polynomial of reasonably low degree (so that the 'unit operations' may be easily carried out) and in which the range of calculation is subdivided into working intervals which are short enough locally to preserve the desired accuracy over the whole calculation.

There are also many types of infinite series expansion in terms of complete sets of orthogonal functions. The most familiar of these are *Fourier series** in terms of cosines, sines, or a mixture of both, and in each case valid within a range of orthogonality of the set of functions involved. Other sets of orthogonal functions which may be used for series expansion in an analogous manner include Legendre polynomials, Chebychev polynomials, Bessel functions, and so on. In principle, each of these expansions can be used in truncated form as a basis for the development of numerical processes in very much the same way as direct or 'simple' polynomial representation. In practice, some of the most useful methods have been developed from truncated Fourier series and series of Chebychev polynomials. Fourier series methods form the basis of harmonic analysis and are particularly useful in handling functions or data with natural periodicities. Chebychev polynomial methods provide the most rapid convergence of all approximations based on special or simple polynomials, and give a special degree of uniformity in the distribution of errors of approximation over the range of application (in contrast with simple polynomials where the error of approximation tends to be smaller in the centre and to increase fairly sharply towards the ends of the range). It may be noted that a similar form of Weierstrass's theorem holds for Fourier and Chebychev representations.

* See Sneddon, *Fourier Series*, in this series

All the methods of approximation which have been described can be collected into the single expression

$$f(x) \simeq a_0\phi_0(x) + a_1\phi_1(x) + \ldots + a_n\phi_n(x),$$

and any such finite approximation may be used as a basis for the development of numerical processes (though only with complete sets of functions can we be sure that the error will be reduced progressively as we include successively more terms of the representation). This representation is an approximate form with error $R_n(x)$ of the exact terminated expansion

$$f(x) = a_0\phi_0(x) + a_1\phi_1(x) + \ldots + a_n\phi_n(x) + R_n(x).$$

In order to emphasize the role of approximation in the processes we shall develop, the symbol $\simeq$ will be used to denote any of these approximate representations, such as polynomial approximation with $\phi_k(x) = x^k$ or Fourier approximation with $\phi_k(x) = \genfrac{}{}{0pt}{}{\cos}{\sin} \pi k x$. Many authors use the symbol $=$ for all purposes, but this can prove confusing to the beginner. In each case the maximum value of $R_n(x)$ (if it can be estimated) in a range of application provides a measure of the accuracy obtainable with that approximation.

A family of numerical processes will be developed in the following chapters from the simple basis of *polynomial approximation*, using the general form

$$f(x) \simeq a_0 + a_1 x + a_2 x^2 + \ldots + a_n x^n,$$

(with appropriately chosen degree n) for known and unknown functions and for sections of tabulated values. Numerical processes based on polynomial representation are relatively easy to derive, simple to apply, and very satisfactory for displaying the principles of numerical approximation; moreover, they require only a very modest mathematical background, and so form an ideal introduction to

NUMBERS AND TABLES OF VALUES

the ideas underlying numerical analysis. In contrast, the development of methods based on approximation by special functions calls for a wider knowledge of analysis; hence in spite of their greater current importance, methods based on approximation by orthogonal functions will be postponed to a later book in the series.

1.3. Numbers, continuous functions, tables of values, and tabulated experimental results

These provide the raw material of numerical analysis.

We shall normally represent numbers in decimal form.*
Thus the number

$$3 \cdot 14159$$

has five *decimals* (or *decimal places*, or *decimal digits*), and six *significant figures*. It is the value of π, and is to be interpreted as being correct to within five units in the sixth decimal place; it is impossible to guarantee greater accuracy without writing additional decimals, and it is an important convention that it should be regarded as dishonest to fill decimal places with digits known to be inaccurate (or, indeed, with digits of unknown accuracy). This represents an approximate value of π which is of course an irrational number, but irrational numbers cannot as such enter numerical analysis since we are forced to approximate to them. In the same way we must approximate all rational numbers which cannot be expressed exactly in the number of decimals with which we choose to work, and for obvious practical reasons this number cannot normally be very large.

Sometimes we have to handle very large or very small numbers, such as Avogadro's number $6 \cdot 023 \times 10^{23}$, or the charge of the electron in e.s.u. $4 \cdot 802 \times 10^{-10}$; and in such cases we usually separate the significant figures (which we write in the form of a workable decimal number of order

---

* This is by no means the only possibility, though it is most convenient for our purposes; modern computing machines use binary form.

unity) from the powers of ten involved, and we carry the two parts of the number separately through the calculation. This may be called *floating point decimal form*, and may also be written 6·023, 23 and 4·802, −10 for these examples. It may be noted that the value for the charge of the electron implies that it is known correctly to about one part in ten thousand (to be precise, 5 parts in 48020).

Continuous functions could in principle be represented numerically by curves showing their value at every point, but for practical purposes the only way in which most functions can be represented accurately and conveniently is by a table of values at specified intervals of the independent variable.* And of course we are only too familiar with such tables of logarithms, sines, cosines, exponentials and so on. In these printed tables each entry represents a true value, correct to within half a unit in the last digit unless someone has blundered, of a precisely defined function. (We must accept the fact that other people make mistakes just as we do ourselves, and quite a lot of mistakes appear in print!) These functions obviously exist between tabulated values, and we shall develop the process of *interpolation* to calculate such intermediate values. However, we cannot tabulate the functions more closely and many functions are tabulated scarcely at all because of the sheer magnitude of the task involved. Thus, a table of five-figure logarithms for the numbers from 10 by intervals of 0·1 to 100 occupies two pages with 900 main entries; a similar table of a function of two variables would take two large books of 900 pages each; one of three variables a room of books; one of four variables (such as the hypergeometric function) a library of more than one and a half million volumes! The evaluation of many functions can be carried out very quickly on fast computing machines.

Tables of experimental results differ from tables of functions

---

* But see Jahnke and Emde, *Tables of Functions*, Dover for some beautifully drawn curves which are extremely effective in displaying the broad types of behaviour of rather complicated functions.

ERRORS IN NUMERICAL ANALYSIS

in that each entry is likely to include experimental error, and although we can place probable bounds on the magnitude of the error we can do nothing to improve the accuracy of individual results without repeating and refining the experimental measurements. There are two kinds of experimental error: systematic and random. *Systematic errors* may be due to a persistent fault in the performance or faulty design of the experiment; they can seldom be detected by direct analysis of the experimental results as each measurement suffers, and they are less a part of the analysis of the results than of experimental design. *Random errors* are often apparent on inspection, specially when the results are presented graphically and show 'scatter'; they can be kept at a low level by careful work but they cannot be eliminated, and they set an obvious limit to the accuracy which can be obtained in calculations using the experimental results.

A common way of approximating a set of experimental results is to plot them on suitable logarithmic scale or other graph paper in an attempt to produce a linear relationship; a 'line of best fit' can then be estimated by eye, and will serve to define a functional approximation to the original results. When the scatter is appreciable, there is some temptation to use one of the *smoothing* techniques so that the adjusted points lie more closely on a smooth curve; however, such devices should be used with great caution, as the raw experimental results provide our only direct link with the physical world and if they do not form an obvious pattern this *may* be because we have taken the wrong measurements. We shall mention more sophisticated methods of *curve fitting* in chapter 5, and it should be noted that there are statistical methods for assessing the significance of variations in experimental measurements.

1.4. Errors and accuracy in numerical analysis

The only useful numerical solution is a correct one—though we must be careful how we interpret the word 'correct'. We can usefully distinguish two causes producing wrong numerical answers: *errors* which arise from the finite character of numerical analysis and are inherent in our processes, and *blunders* caused by our own carelessness or the failure of machine aids.

Inevitably we must work with a limited number of significant figures, because our machines have limited capacity and indeed we have even more limited capacity ourselves in practice! Few numbers will be small integers, and there will usually be a *rounding off error* produced by the rejection of all digits after a selected decimal place. Our procedure for rounding off will be to preserve unchanged the last digit retained on discarding fewer than five units in the next place, to increase it by one on discarding more than five units, and to round off to the nearest *even* number on discarding exactly five units. Some authors always round five units in one direction (usually up), but our procedure preserves a randomness and so increases the chance of mutual cancelling of rounding errors during arithmetical operations. Thus, on rounding to three decimals,

$$0\cdot 5772 \to 0\cdot 577, \quad 3\cdot 14159 \to 3\cdot 142,$$
$$0\cdot 0625 \to 0\cdot 062, \quad 0\cdot 1875 \to 0\cdot 188.$$

This procedure normally ensures that the rounding error will not exceed half a unit in the last digit retained.

Rounding errors can accumulate in arithmetical operations, and it may be necessary to carry an extra decimal digit, otherwise unnecessary, rounding off at the end of a chain of operations. The probable error will be considerably less than the maximum error, and can be estimated statistically. For example, the maximum error in the sum of n rounded-off values is $\pm \frac{1}{2} n$ units in the last digit and occurs only when each value has the greatest possible rounding error and all reinforce; in practice the distribution of rounding errors is likely to be almost random and for large n the probable error in the sum is likely to be less than $\pm \frac{1}{2} \sqrt{n/3}$, with error about $\pm \sqrt{n/3}$ in roughly one entry in twenty.

ERRORS IN NUMERICAL ANALYSIS

Frequently we obtain finite numerical processes by suitably truncating an appropriate infinite series. This neglect of higher terms introduces a *truncation error*, which can be kept small by retaining a sufficient number of terms (though with rapidly increasing labour as the number of terms increases). Truncation errors will be present in most processes other than those of arithmetic.

Blunders are lapses from accurate computation due to carelessness of the computer or failure of the machine. In principle they are avoidable, but in practice they are almost impossible to eliminate; hence we must be constantly on guard, and the checking and rechecking of calculations must become a part of our habitual practice as computers. For example, in adding a column of figures we should always do it twice, once from the bottom and once from the top; if the sums do not agree we must repeat until two successive checks *do* give the same answer. Whenever possible, checks should be made using different methods of calculation, because if we have made a mistake once we are quite likely to make it again in a repeat calculation; careful computational design will often produce formulae which are not used directly in the calculation but must nevertheless be satisfied, and these provide valuable checks. We should always finish a problem with two consecutive checks that agree.

Beginners often try to laugh off this emphasis on blunders, but they learn how easy it is: to transpose digits, taking 7239 in place of 7329; to misread repeated digits, taking 72239 in place of 72339; to misread tables; and to overlook signs, especially near sign changes. They find that after identifying a mistake they will make a wrong correction. These and other blunders can be kept to a minimum only when the computer learns to work with an easy rhythm, to write his figures clearly and to lay out his computation in an orderly fashion. And he will discover his remaining blunders quickly if he checks the work stage by stage, avoiding checks by repetition; finally he will be confident of his answer only when he has completed an over-all check.

Inexperienced computers often carry too many—sometimes far too many—decimals through their calculations from lack of confidence, and from a vague idea that they may increase the accuracy of their work. We must understand clearly that although our model theories in science are capable of unlimited accuracy mathematically, they are based on approximate physical assumptions and are directed towards approximate physical measurements. In many ways the bridge of mathematics in between is expendable, and there is certainly no merit to be gained from unreasonable pursuit of accuracy in our model theories. Indeed the very striving for excessive accuracy defeats its own purpose; for the computer, burdened by excessive length in his calculations, will quickly tire and err. We shall do well to recognize as the mark of the experienced computer the fact that he plans his problem so that at each stage he carries just that accuracy required locally to produce a final answer as accurate as the experimental measurements and no more so.

1.5. Effective numerical computation

The criterion by which we must judge numerical processes is *effectiveness*. This involves mathematical considerations such as accuracy and availability of independent checks, but also what may best be described as economic considerations such as the time required to obtain specified accuracy. When there is a long run of similar calculations we shall spend ample time in reducing the problem to the best possible form for computation, so that each case may be handled quickly and accurately with an over-all saving in time and effort. But if we have a single case to compute we shall use any already-familiar process, and the calculation will be finished in less time than a preparatory reduction. We shall, of course, use mathematical tables, slide rules, hand machines, and above all else high-speed electronic computers whenever they will be helpful.

EXERCISES

A glance at some of the larger texts will show that there are very many processes available for each type of calculation, such as interpolation or integration, for example. No doubt the beginner would wish to see one process of each type labelled best, but this cannot be done. *There are no 'best processes'; the most effective processes will always depend on the particular problem in hand.* There is great scope for the use of *judgment* in numerical analysis in such matters as the selection of an appropriate process, the detailed design of a calculation, the choice of working interval, number of working digits, and so on. Indeed it is this demand on the capacity for skilled judgment by the computer that provides the challenge and much of the interest of numerical work.

In this short book and a companion volume on *Numerical Solution of Equations* we shall discuss the kinds of way in which numerical processes can be constructed and put to use. We shall introduce few out of the vast array of available formulae, but as the beginner enlarges his experience he will learn how to select the most suitable from among many processes that might be used on a given problem.

EXERCISES ON CHAPTER ONE

1. State the number of decimal places and of significant figures in each of the following numbers, and (approximately) the accuracy to which they are presented: (i) 980.665, (ii) 9.11×10^{-28}, (iii) 2.9978×10^{10}, (iv) 1.20×10^{-2}.

2. Compare the result of rounding off 1.37455 digit by digit until only two decimals remain, and rounding off to two decimals in a single operation.

3. Evaluate and write in appropriate form, giving an estimate of the maximum error in each answer:
  (i) $1.73 - 2.16 + 0.08 + 1.00 - 2.23 - 0.97 + 3.02$,
  (ii) 9.37×8.91,
  (iii) $7.137/0.279$,
  (iv) $2.9132 + 15.164 - 0.31725$.

4. Compare the sum and greatest rounding error when the following terms are added and then rounded to six decimals with those obtained when the terms are first rounded off and then added:

·5833333 + ·1000000 + ·0083333 + ·0011905 + ·0002226 +
    + ·0000493 + ·0000122 + ·0000033.

5. Obtain linear, quadratic and cubic polynomial approximations for e^x near $x=0$ by truncating the Taylor series appropriately, and use the remainder term to estimate (to the nearest 0·1) the range over which each polynomial can be used if the error of approximation is not to exceed 0·005 (i.e. the approximation is to be correct to two decimals).

CHAPTER TWO

Finite Differences

Before we can develop particular processes we must investigate the operations that can be carried out directly on a table of values. Continuous functions must normally be represented numerically by tables of values at specified intervals of the argument, and the limit processes of infinitesimal analysis can no longer be applied to such tabulated functions but must be replaced by equivalent finite processes.

2.1. Tables of values

Most functions in common use are tabulated, and these tables of values provide one of our sources of raw material. Any scientist should become familiar with selected sets of tables: for example Comrie, *Chambers's Shorter Six-figure Mathematical Tables* is good for ordinary functions, and Jahnke & Emde, *Tables of Functions* is useful for some less common ones. To illustrate table notation we may take as typical Chambers's table of ordinary logarithms 1,000 (1) 10,000, 6D; the table gives in a rectangular array the mantissae correct to six decimals (denoted by 6D) for the numbers from 1,000 to 10,000 (the endpoints and the tabular interval 1 in the argument are denoted by 1,000 (1) 10,000); decimal points and characteristics are suppressed as they convey no additional information of value, and their inclusion would make the table harder to work from. Tabulated values correct to six significant figures are denoted by 6S. Another illustration is provided by Chambers's table for e^x given at values of the argument 0·000 (·001)

3·000 (·01) 10·00 (·1) 50·0 to between six and seven significant figures; the entries are in decimal form up to $x=15·0$, and thereafter in floating point decimal form 6D, 7S so that the entry for $e^{15}=3,269,017$ is printed as 3·269017, 6.

Some functions change rapidly in isolated regions of the argument and slowly elsewhere, and such functions can best be tabulated by using different tabular intervals in different parts of the range according to the local behaviour of the function. The resulting tables are rather more difficult to work from than those with uniform interval, and the great majority of the tables in common use are based on a constant tabular interval in the argument, at least over substantial regions (e.g. e^x above). A section of such a table, say for $f(x)=1/x$, might be written

x	3·0	3·1	3·2	3·3	3·4	3·5	3·6	3·7
$1/x$	·33333	·32258	·31250	·30303	·29412	·28571	·27778	·27027

This table cannot represent the nominal function $1/x$ exactly; for there will be rounding errors in most entries, and indeed the tabulated function is not formally defined at all for intermediate values of x. In this case we know very well that the tabulated values provide a good approximation to $1/x$, and that intermediate values can be estimated by fitting a polynomial (Weierstrass's theorem!) and we might anticipate (or hope?) that a low degree polynomial would be adequate because of the slow variation of $1/x$ in this interval. But what if we are merely presented with a table of values? In principle this could represent some pathological function of pure mathematics, and even if Weierstrass's theorem should be applicable a very high degree polynomial might be required for matching; or it might represent some perfectly well-behaved periodic function with natural wavelength considerably shorter than the range of the table. Thus, table 1 gives values for $\sin 2\pi x$ at three different working intervals $h=0·9$, $1·0$, and $1·05$ for the variable x; it is 'perfectly clear' that each of these

Table 1

x		0	h	$2h$	$3h$	$4h$	$5h$
	0·9	0	−·588	−·951	−·951	−·588	0
$\sin 2\pi x$, $h=$	1·0	0	0	0	0	0	0
	1·05	0	·309	·588	·809	·951	1·000

three sets of values defines a 'good smooth function', but unfortunately none of these 'good smooth functions' bears the slightest useful relationship to the original function $\sin 2\pi x$! In each of these (rather artificial) cases the working interval is so large relative to the periodic length of the function that the main periodicity has entirely escaped representation in the table. An even worse example is provided by the function $\sin(2\pi/x)$, for in this case as x decreases, no matter how small a working interval we may select it will soon become impossibly large relative to the continuously shortening periodic length of the function. Clearly there can be no certain way of knowing whether a given table provides a proper representation of the (unknown) function which it is supposed to represent. The interval adopted in printed tables will always be short enough to display fully the essential character of the function, and it is important that we should always adopt the same convention ourselves when giving tables of values; and it should be noted that this will usually mean using different intervals in different parts of the range according to the local behaviour of the function. We shall assume in working from tables that the higher order derivatives of the function are small enough for the function to behave 'reasonably smoothly' over the length of the tabular interval.

2.2. Finite differences

The simplest operation on a table of values, obtained either for a known function or by calculation, is to find the *difference* between each pair of tabular values. The *first differences* are found by subtracting each value from its

FINITE DIFFERENCES

successor in the table, *second differences* by repeating a similar operation on the first differences, and so on for higher orders; these together comprise the *finite differences* of the table. The customary form for tabulation is:

x	$f(x) = 1/x$	*first-*	*second- differences*	*third-*
3·0	·33333			
		−1075		
3·1	·32258		67	
		−1008		−6
3·2	·31250		61	
		−947		−5
3·3	·30303		56	
		−891		−6
3·4	·29412		50	
		−841		−2
3·5	·28571		48	
		−793		−6
3·6	·27778		42	
		−751		
3·7	·27027			

Note: (i) that odd-order differences are written between, and even-order differences on the tabular lines; (ii) that decimal points and leading zeros are omitted in the differences in order to simplify the table; (iii) that we must very definitely *not* omit the signs; (iv) that each difference must be multiplied by 10^{-5} before use.

Two features stand out in this table. (1) Differences of increasing order decrease quite rapidly in magnitude. We shall show in §2.4 that this type of behaviour is typical of polynomials and of functions or tables which are well-fitted for polynomial approximation. (2) The third differences show marked irregularity. If, however, we tabulate $1/x$ to six decimals (6D) we find that the retention of the additional decimal has produced fairly regular third differences, but markedly irregular fourth differences:

2.2 FINITE DIFFERENCES

x	$1/x$	first-	second-	third-	fourth-
				differences	
3·0	·333333				
		−10752			
3·1	·322581		671		
		−10081		−60	
3·2	·312500		611		7
		−9470		−53	
3·3	·303030		558		3
		−8912		−50	
3·4	·294118		508		10
		−8404		−40	
3·5	·285714		468		0
		−7936		−40	
3·6	·277778		428		
		−7508			
3·7	·270270				

This behaviour can be seen to arise from rounding errors; even though individual rounding errors in $f(x)$ are $\leq \frac{1}{2}$ in the last decimal digit it is clear that these errors are likely to accumulate, and be increasingly amplified in successively higher differences. The *greatest* error that can be obtained is, approximately,

tabular error				differences			
	1st	2nd	3rd	4th	5th	6th	7th
$+\frac{1}{2}$							
	−1						
$-\frac{1}{2}$		+2					
	+1		−4				
$+\frac{1}{2}$		−2		+8			
	−1		+4		−16		
$-\frac{1}{2}$		+2		−8		+32	
	+1		−4		+16		−64
$+\frac{1}{2}$		−2		+8		−32	
	−1		+4		−16		
$-\frac{1}{2}$		+2		−8			
	+1		−4				
$+\frac{1}{2}$		−2					
	−1						
$-\frac{1}{2}$							

FINITE DIFFERENCES

These are *maximum* errors, and the likely errors will be smaller as the rounding error will not usually alternate in sign; Comrie* has given a working criterion for the *expected* fluctuations which will arise from round-off as

order of difference	1	2	3	4	5	6
expected limits of error	±1	±2	±3	±6	±12	±22,

and we must learn to tolerate small irregularities of this magnitude in our tables of differences. This level of fluctuations has been called the 'noise level' of a difference table.

The magnitude of the finite differences obviously depends on the length of the tabular interval.

Although tables of values play a more important role in hand calculation than with high-speed machines, they are generally important for the expression of experimental results and of the results of a previous calculation. They are seldom used for the description of specified functions with fast machines, as it is usually much quicker to use a programme sub-routine to evaluate a function where it is needed than to feed in a table of numbers.

2.3. Notation for finite differences

There are three distinct notations for the single set of finite differences of a function $f(x)$ tabulated at constant interval h in the independent variable x. If we write $x_n \equiv x_0 + nh$ for the nth tabular point, and $f_n \equiv f(x_n)$, then we define

(i) The *forward difference* at x_i is $f_{i+1} - f_i$, written Δf_i. Thus

$$\Delta f(x_0) = f(x_0 + h) - f(x_0) = f_1 - f_0,$$

and higher differences $\Delta^2 f_0 = \Delta(\Delta f_0) = \Delta f_1 - \Delta f_0 = f_2 - 2f_1 + f_0$, $\Delta^3 f_2 = \Delta(\Delta^2 f_2) = f_5 - 3f_4 + 3f_3 - f_2$, etc., can be calculated either by differencing lower differences or from the tabulated values. Note that

* See introduction to *Chamber's Six-figure Mathematical Tables*, Vol. II.

2.3 NOTATION FOR FINITE DIFFERENCES

$$\Delta^k f_i = f_{k+i} - \binom{k}{1} f_{k+i-1} + \binom{k}{2} f_{k+i-2} - \ldots + (-1)^k f_i,$$

where $\binom{k}{r}$ are the binomial coefficients.

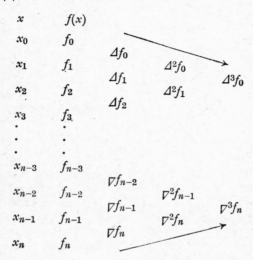

(ii) The *backward difference* at x_i is $f_i - f_{i-1}$, written ∇f_i. Thus, for example,

$$\nabla^2 f_n = \nabla(f_n - f_{n-1}) = f_n - 2f_{n-1} + f_{n-2}.$$

The forward and backward differences represent in aggregate precisely the same set of numbers (and $\nabla f_i \equiv \Delta f_{i-1}$), but forward differences are specially useful near the start of a table as they involve only the tabular values *below* the 'subscript level' x_i, and backward differences are useful near the finish of a table where we do not have the forward tabular values to obtain all the necessary forward differences.

(iii) The *central difference* at x_i is $f_{i+1/2} - f_{i-1/2}$, written

FINITE DIFFERENCES

δf_i. Thus $\delta f_{i+1/2} = f_{i+1} - f_i = \Delta f_i = \nabla f_{i+1}$, etc. This is the normal notation for use away from the ends of a table, where there are plenty of tabular values above and below the level x_i.

$$
\begin{array}{cccccc}
f_{i-2} & & & & & \\
 & \delta f_{i-3/2} & & & & \\
f_{i-1} & & \delta^2 f_{i-1} & & & \\
 & \delta f_{i-1/2} & & \delta^3 f_{i-1/2} & & \\
f_i & \longrightarrow & \delta^2 f_i & \longrightarrow & \delta^4 f_i \\
 & \delta f_{i+1/2} & & \delta^3 f_{i+1/2} & & \\
f_{i+1} & & \delta^2 f_{i+1} & & \delta^4 f_{i+1} \\
 & \delta f_{i+3/2} & & \delta^3 f_{i+3/2} & & \\
f_{i+2} & & \delta^2 f_{i+2} & & & \\
 & \delta f_{i+5/2} & & & & \\
f_{i+3} & & & & &
\end{array}
$$

Note that $\Delta^n f_i$ is the nth forward difference and lies on the downwards diagonal from the tabular entry f_i, while the nth backward difference $\nabla^n f_i$ lies on the upwards diagonal originating at f_i. The nth order central difference $\delta^n f_i$ lies on the horizontal level x_i; thus odd central differences such as $\delta f_{1/2}$ lie on half tabular lines and even ones like $\delta^2 f_{-1}$ on tabular lines.

2.4. Finite differences of a polynomial

The finite differences of a general polynomial expression can be found only in terms of a specified tabular interval h.

The finite difference of a constant over any interval is, of course, zero.

The successive forward differences of the power x^k are

$$\Delta x^k = (x+h)^k - x^k = khx^{k-1} + \tfrac{1}{2}k(k-1)h^2 x^{k-2} + \ldots$$
$$+ kh^{k-1}x + h^k,$$
$$\Delta^2 x^k = \Delta(\Delta x^k) = k(k-1)h^2 x^{k-2} + k(k-1)(k-2)h^3 x^{k-3}$$
$$+ \ldots + (2^k - 2)h^k,$$

2.4 FINITE DIFFERENCES OF A POLYNOMIAL

$$\Delta^i x^k = k(k-1)\ldots(k-i+1)h^i x^{k-i} + \tfrac{1}{2}ik(k-1)\ldots$$
$$(k-i)h^{i+1}x^{k-i-1} + \ldots,$$

. . . .

$$\Delta^k x^k = k!\, h^k.$$

All higher differences $\Delta^{k+i}x^k$, $i>0$ are zero. Thus the ith difference of x^k is a polynomial of degree $k-i$ for $0<i\leqslant k$, and is zero for $i>k$.

The finite differences of an nth degree polynomial

$$f(x) = a_0 + a_1 x + a_2 x^2 + \ldots + a_n x^n$$

follow, and it can be seen that $\Delta^i f(x)$ is a polynomial of lower degree $n-i$ if $0<i\leqslant n$, and is zero if $i>n$. In particular,

$$\Delta^n(a_0 + a_1 x + \ldots + a_n x^n) = a_n n!\, h^n,$$
$$\Delta^{n+1}(a_0 + a_1 x + \ldots + a_n x^n) = 0.$$

Thus the nth difference of a polynomial of degree n is a constant proportional to h^n, and higher differences are zero.

Example 1. Derive finite difference tables for the polynomial $f(x) = 2 - 3x^2 + x^3$ with the two working intervals $h = 0.5$ and $h = 0.1$.

These tables can easily be obtained as:

x	$f(x)$	Δf	$\Delta^2 f$	$\Delta^3 f$	$\Delta^4 f$
0	2·000				
		−625			
·5	1·375		−750		
		−1375		750	
1·0	0		0		0
		−1375		750	
1·5	−1·375		+750		0
		−625		750	
2·0	−2·000		+1500		·
		+875		·	
2·5	−1·125		·		·
·	·		·	·	·

FINITE DIFFERENCES

x	$f(x)$	Δf	$\Delta^2 f$	$\Delta^3 f$	$\Delta^4 f$
0	2·000				
		−29			
·1	1·971		−54		
		−83		6	
·2	1·888		−48		0
		−131		6	
·3	1·757		−42		0
		−173		6	
·4	1·584		−36		·
		−209		·	
·5	1·375		·		·

Note: (i) that in each case third differences are constant and equal to $3! \, h^3$, and fourth differences are zero; (ii) that the differences depend on h and higher differences are systematically small only when h is small; (iii) that if tabular values of $f(x)$ are rounded off there will be small irregularities in fourth and higher differences.

A comparison of the second table of example 1 with the table for $1/x$ on page 18 suggests that after allowing for round-off fluctuation we might very well adopt a cubic approximation for $1/x$ in the range $3 \cdot 0 \leqslant x \leqslant 3 \cdot 7$. Indeed we shall find that *whenever the higher differences of a table become increasingly small, the function represented by the table is well-fitted for polynomial approximation.* Large differences may be reduced by reduction of the working interval h, but in some cases approximation by reasonably low-degree polynomials will be possible only for inconveniently small h and other types of approximation may be more effective. Thus, it will be found that the values in table 1 are ill-suited for polynomial approximation and will remain so unless the interval is reduced considerably; this is hardly surprising for a sinusoidal function of wavelength small relative to the range of x, and we might expect that harmonic approximation would be better.

2.5. Detection of mistakes by differencing

Rounding errors cause only small fluctuations in differences, but blunders will produce much larger upsets. Hence a standard and invaluable method for detecting mistakes in any table is to form differences and inspect those of higher order for irregularity. An error of one unit in $f(x)$ (or in a difference), and a pair of adjacent errors propagate through the differences as follows:

```
0                           1        0            1
    0       0       1          -6        0      1     -3
0       0       1      -5           1      -2
    0       1      -4      15        1     -1       2
        1      -3      10               0      0
1      -2       6     -20        1     -1       2
   -1       3     -10              -1      2
0       1      -4      15        0      1     -3
    0      -1       5                0     -1
0       0       1      -6        0      0       1 etc.
    0       0      -1                0
0       0       0       1 etc.  0
```

Note how in each case the fanwise spread of the error locates its position, and that a single error grows on propagation as the binomial coefficients of increasing order. Also that the sum of the errors in each column crossing an error fan is zero. Hence, if we can establish the general trend of the affected differences in a column from those above and below an error fan, we can infer what these differences ought to be, i.e. what correction is needed. In the simplest case of approximately constant differences the errors are obtained by a comparison of mean and actual values of the differences.

Thus we detect errors in a table of values from the fact that a single error ϵ produces maximum errors

	ϵ	2ϵ	3ϵ	6ϵ	10ϵ	20ϵ	
in the	1st	2nd	3rd	4th	5th	6th	dif-

FINITE DIFFERENCES

ferences, respectively. *This is a powerful check and should always be applied to any table produced by calculation.*

Example 2. Check the following table for $\log_e x$.

x	$\log x$	differences			recalculated differences		
1·0	·00000						
		9531			9531		
1·1	·09531		−830			−830	
		8701		133	8701		133
1·2	·18232		−697			−697	−29
		8004		131	8004	104	
1·3	26236		−566			−593	−23
		7438		0	7411	81	
1·4	·33674		−566			−512	−14
		6872		148	6899	67	
1·5	·40546		−418			−445	−13
		6454		27	6454	54	
1·6	·47000		−391			−391	50
		6063		104	6063	104	
1·7	·53063		−287			−287	−187
		5776		−83	5776	−83	
1·8	·58779		−370			−370	177
		5406		94	5406	94	
1·9	·64185		−276			−276	
		5130			5130		
2·0	·69315						

The third differences indicate mistakes, probably more than one. Apparently $\Delta^3 f_{1.1}$ and $\Delta^3 f_{1.3}$ are high, $\Delta^3 f_{1.2}$ and $\Delta^3 f_{1.4}$ low, suggesting too large an entry for $\log_e 1.4$; the necessary reduction lies between 20 and 30 (as may be verified by redifferencing), and this suggests very strongly that the digits 74 have been interchanged from a correct value $\log_e 1.4 = .33647$. The differences have been recalculated after this correction, and the surviving error fan is apparently centred on $\Delta f_{1.7}$; a reduction

2.5 DETECTION OF MISTAKES

of about 60 in $\Delta f_{1.7}$ produces reasonably uniform fourth differences, and on checking we find that $\Delta f_{1.7}$ should in fact be 5716.

Note: (i) it should be verified that the fourth differences are now as uniform as can be expected; (ii) a convenient check can be made by comparing the sum of entries in any column of differences with the difference between the first and last entries of the preceding column—this check discloses the existence but not the position of the mistake; (iii) differencing will uncover mistakes but not systematic errors.

Problem 1. Tabulate $x^3 - 2x^2 + 3x - 1$ to 4D for $x = 0(\cdot 01)0\cdot 1$. We could write a programme to calculate and print out the

x	$f(x)$	∇	∇^2	∇^3	check		
$-\cdot 02$	$-1\cdot 060808$				$-1\cdot 0608$		
		30607				306	
$-\cdot 01$	$-1\cdot 030201$		-406		$-1\cdot 0302$		-4
		30201		6		302	
$\cdot 00$	-1		-400		-1		-4
		29801		6		298	
$\cdot 01$	$-0\cdot 970199$		-394		$-0\cdot 9702$		-4
		29407		6		294	
$\cdot 02$	$-\cdot 940792$		-388		$-\cdot 9408$		-4
		29019		6		290	
$\cdot 03$	$-\cdot 911773$		-382		$-\cdot 9118$		-3
		28637		6		287	
$\cdot 04$	$-\cdot 883136$		-376		$-\cdot 8831$		-5
		28261		6		282	
$\cdot 05$	$-\cdot 854875$		-370		$-\cdot 8549$		-3
		27891		6		279	
$\cdot 06$	$-\cdot 826984$		-364		$-\cdot 8270$		-4
		27527		6		275	
$\cdot 07$	$-\cdot 799457$		-358		$-\cdot 7995$		-3
		27169		6		272	
$\cdot 08$	$-\cdot 772288$		-352		$-\cdot 7723$		-4
		26817		6		268	
$\cdot 09$	$-\cdot 745471$		-346		$-\cdot 7455$		-3
		26471		6		265	
$\cdot 10$	$-\cdot 719000$		-340		$-\cdot 7190$		

FINITE DIFFERENCES

value of the polynomial at each successive value of x if we were using a high-speed machine, but this approach is unnecessarily laborious if we are working by hand. For we can build up successive values of the polynomial by accumulation of differences if we start from a few calculated values and use the fact that $\nabla^3 f(x) = \text{const}$.

The section of table above the stepped line has been obtained by summing the terms of the polynomial, using values symmetrically disposed about $x = 0$ for simpler calculation. The value $f(\cdot 03)$ is then obtained from the sequence of operations: $\nabla^3 f(\cdot 03) = \cdot 000006$, $\nabla^2 f(\cdot 03) = \nabla^2 f(\cdot 02) + \nabla^3 f(\cdot 03)$, $\nabla f(\cdot 03) = \nabla f(\cdot 02) + \nabla^2 f(\cdot 03)$, $f(\cdot 03) = f(\cdot 02) + \nabla f(\cdot 03)$. This is indicated by arrows.

Note: (i) that even though the third difference is considerably smaller than the rounding error in four decimals it cannot be neglected because this would constitute a *systematic* error which would accumulate; (ii) that there is an easy check by substitution at $x = \cdot 1$; (iii) that the *rounded* values (which make up the final solution) have been checked by differencing; (iv) that backward differences are used as these depend on values already calculated.

2.6. Divided differences

Suppose that we have to deal with a function $f(x)$ tabulated at values x_i of the argument x, where the tabular intervals are *unequal*. This calls for a more general type of finite difference taking account both of the change in tabular value $f_{i+1} - f_i$ and of the length of interval $x_{i+1} - x_i$; the obvious choice is the ratio $(f_{i+1} - f_i)/(x_{i+1} - x_i)$ which gives a measure of the change in tabular value per unit interval. We define the *first divided difference* as

$$[x_i x_{i+1}] = \frac{f(x_{i+1}) - f(x_i)}{x_{i+1} - x_i}.$$

Similarly the first divided difference for any pair of tabular values f_i and f_j (not necessarily consecutive or positively ordered) is

2.6 DIVIDED DIFFERENCES

$$[x_i x_j] = \frac{f(x_j) - f(x_i)}{x_j - x_i} = [x_j x_i].$$

The *second divided difference* $[x_i x_{i+1} x_{i+2}]$ of the tabular values $f(x_i)$, $f(x_{i+1})$, $f(x_{i+2})$ is defined as

$$[x_i x_{i+1} x_{i+2}] = \frac{[x_{i+1} x_{i+2}] - [x_i x_{i+1}]}{x_{i+2} - x_i},$$

the *third divided difference* as

$$[x_i x_{i+1} x_{i+2} x_{i+3}] = \frac{[x_{i+1} x_{i+2} x_{i+3}] - [x_i x_{i+1} x_{i+2}]}{x_{i+3} - x_i},$$

and so on for higher orders.

The divided difference of a set of adjacent tabular values occupies a position in the difference table at the apex of the triangle of diagonals based on the tabular values, and its value and position are entirely independent of the ordering of terms in the divided difference. This may be seen if the divided differences are written symmetrically in terms of the tabular values; for example,

$$\begin{aligned}
&[x_0 x_1 x_2] \\
&= \frac{[x_1 x_2] - [x_0 x_1]}{x_2 - x_0} = \frac{1}{x_2 - x_0} \left\{ \frac{f(x_2) - f(x_1)}{x_2 - x_1} - \frac{f(x_1) - f(x_0)}{x_1 - x_0} \right\} \\
&= \frac{f(x_0)}{(x_0 - x_1)(x_0 - x_2)} + \frac{f(x_1)}{(x_1 - x_0)(x_1 - x_2)} + \frac{f(x_2)}{(x_2 - x_0)(x_2 - x_1)}
\end{aligned}$$

Similar results hold for differences of all orders (proof by induction). It can be seen from the symmetric form of these expressions that the ordering of the arguments is not significant, and that the value of the divided difference depends only upon *which* arguments are present; thus, for example, $[x_0 x_3 x_1 x_2] = [x_0 x_1 x_2 x_3]$.

The nth divided differences of a polynomial of degree n are constant and higher differences zero; for if $f(x) = x^n$,

FINITE DIFFERENCES

$$[xx_0] = [x_0x] = \frac{x^n - x_0^n}{x - x_0} = x^{n-1} + x^{n-2}x_0 + \ldots + x_0^{n-1},$$

so that the first divided difference is a polynomial of degree $n-1$, and the result follows by repeated differencing.

Example 3. Draw up a table of divided differences for $f(x) = x^3$ tabulated at $x = 0, 5, 9, 12, 14, 15$.

x	$f(x)$		divided differences		
0	0				
		25			
5	125		14		
		151		1	
9	729		26		0
		333		1	
12	1728		35		0
		508		1	
14	2744		41		
		631			
15	3375				

The divided differences for a table of *constant tabular interval* h are simply related to the finite differences introduced in §2.3; in particular, the nth divided difference of $n+1$ consecutive tabular values is equal to $\Delta^n f/(n!\,h^n)$ (and it should be noted that $\Delta^n f_i = \nabla^n f_{i+n} = \delta^n f_{i+n/2}$, as these are alternative names for a single difference). The proof is by induction. Suppose that $[x_i x_{i+1} x_{i+2} \ldots x_{i+n}] = \Delta^n f_i/(n!\,h^n)$ for all i; then

$$[x_i x_{i+1} \ldots x_{i+n+1}] = \frac{[x_{i+1} x_{i+2} \ldots x_{i+n+1}] - [x_i x_{i+1} \ldots x_{i+n}]}{(x_{i+n+1} - x_i)}$$
$$= \{\Delta^n f_{i+1}/(n!\,h^n) - \Delta^n f_i/(n!\,h^n)\}/(n+1)h$$
$$= \Delta^{n+1} f_i/\{(n+1)!\,h^{n+1}\}.$$

The result is clearly true for first-order divided differences, and hence it follows that it is true generally.

Problem 2. Evaluate $(x - x_0)(x - x_1) \ldots (x - x_n)[xx_0x_1x_2 \ldots x_n]$ at $x = x_i$, $i = 0, 1, \ldots n$.

EXERCISES

When two of the arguments in a divided difference are equal the formal definition reduces to $0/0$ and further consideration is necessary. However,

$(x-x_i)[xx_0 \ldots x_i \ldots x_n]$
$= (x-x_i)[xx_0 \ldots x_{i-1}x_{i+1} \ldots x_n x_i]$
$= (x-x_i) \dfrac{[x_0 \ldots x_{i-1}x_{i+1} \ldots x_n x_i] - [xx_0 \ldots x_{i-1}x_{i+1} \ldots x_n]}{x_i - x}$
$= [x_0 \ldots x_{i-1}xx_{i+1} \ldots x_n] - [x_0 \ldots x_{i-1}x_i x_{i+1} \ldots x_n],$

and hence as $x \to x_i$, $(x-x_i)[xx_0 \ldots x_i \ldots x_n] \to 0$.

EXERCISES ON CHAPTER TWO

Little attempt has been made to provide straightforward exercises on this chapter as such exercises can be constructed with ease by anyone possessing a set of tables. Readers are urged to work such exercises so that they may become familiar with the various processes and the ideas underlying them.

1. Show that: (i) $\Delta \nabla f_i = \nabla \Delta f_i = \delta^2 f_i$,
(ii) $\Delta^n f_i = \nabla^n f_{i+n} = \delta^n f_{i+n/2}$.

(Note that in each case these are merely alternative ways of denoting the number occupying a particular position in the difference table.)

2. Show that: (i) $\Delta(f_i g_i) = f_i \Delta g_i + g_{i+1} \Delta f_i$, (ii) $\Delta(f_i^2) = (f_i + f_{i+1}) \Delta f_i$,
(iii) $\Delta(f_i/g_i) = (g_i \Delta f_i - f_i \Delta g_i)/(g_i g_{i+1})$, (iv) $\Delta(1/f_i) = -\Delta f_i/(f_i f_{i+1})$.

3. Show that: (i) $\Delta^n (1/x) = \dfrac{(-1)^n n! \, h^n}{x(x+h) \ldots (x+nh)}$,

(ii) $\Delta^n e^{ax} = (e^{ah}-1)^n e^{ax}$,

where h is the length of the working tabular interval.

4. Find the polynomial of lowest degree which fits the data:

x	0	1	2	3	4	5
y	75	150	204	241	265	280.

5. Find which tabular value is wrong in the following table and correct it.

x	$f(x)$	x	$f(x)$	x	$f(x)$
1·0	·340735	1·5	1·142289	2·0	1·793724
1·1	·500429	1·6	1·293533	2·1	1·883902
1·2	·662671	1·7	1·435910	2·2	1·956910
1·3	·825289	1·8	1·567997	2·3	2·011438
1·4	·986092	1·9	1·687869	2·4	2·046371

CHAPTER THREE

Interpolation

We assume here that tables of values provide good representations of well-behaved functions. Hence when we seek to *interpolate*, or estimate non-tabular values of a function, we take for granted that the function will behave smoothly between tabular points and that reasonable approximations can be obtained as polynomials of moderate degree. A polynomial of degree k will have constant kth differences; hence we can judge the suitability of a table for polynomial representation by the behaviour of its higher differences, and *when the kth order differences are nearly constant the function behaves in this range very much as a polynomial of degree* k.

3.1. Linear interpolation

When the first differences of a table are nearly constant the function which it represents must behave very nearly linearly and can be approximated in each interval by a straight line joining adjacent tabular points; this is the basis of the familiar process of *linear interpolation* used almost universally in tables of logarithms and trigonometrical functions. Thus in the interval $x_0 < x < x_1$,

$$f(x) \simeq f(x_0) + \frac{x - x_0}{x_1 - x_0}\{f(x_1) - f(x_0)\}.$$

When we are using tables of constant interval h it will be convenient to write $x - x_0 = \theta(x_1 - x_0) = \theta h$, where $0 < \theta < 1$, and then in the interval (x_i, x_{i+1})

$$f(x_i + \theta h) \simeq f_i + \theta(f_{i+1} - f_i) \equiv (1-\theta)f_i + \theta f_{i+1};$$

or, using the various (entirely equivalent) expressions for the finite difference,

3.1 LINEAR INTERPOLATION

$$f(x_i + \theta h) \simeq f_i + \theta \Delta f_i \equiv f_i + \theta \nabla f_{i+1} \equiv f_i + \theta \delta f_{i+\frac{1}{2}}.$$

Linear interpolation provides a simple process but one that can only be successful when the first differences vary very slowly, and this implies close tabulation to an extent that will be practicable only for very commonly used functions (e.g. 'log' tables). In most tables we need interpolating polynomials of higher degree and we can no longer base an estimate on just *two* tabular values.

Example 4. The following extract is from Knott's 4D log tables.

x	0	1	2	3	4	5	6	7
10	0000	0043	0086	0128	0170	0212	0253	0294
11	0414	0453	0492	0531	0569	0607	0645	0682
12	0792	0828	0864	0899	0934			
13	1139							

8	9	1	2	3	4	5	6	7	8	9
					proportional parts					
0334	0374	4	8	12	17	21	25	29	33	37
0719	0755	4	8	11	15	19	23	26	30	34

Differencing the values of the first row:

x	$f(x)$	Δ	Δ^2
100	0000		
		43	
101	0043		0
		43	
102	0086		−1
		42	
103	0128		0
		42	
104	0170		0
		42	
105	0212		−1
		41	
106	0253		0
		41	
107	0294		−1
		40	
108	0334		etc.

INTERPOLATION

The differences are locally almost constant and the table is suitable for linear interpolation: thus

$$f(1033) = 0128 + 3 \times 42/10 = 0141.$$

The subsidiary table of *proportional parts* shows tenths of the mean difference between tabular values for a row of entries; it should be noted that the values printed are a compromise, and that the proportional parts between 100 and 102 should be 4 9 13 17 22 26 30 34 39 with those between 107 and 110 as 4 8 12 16 20 24 28 32 36, and a single table can serve the range (100, 110) only at the cost of some loss in accuracy (very nearly of the fourth decimal digit!). Good tables do not use averaged proportional parts because of the consequent loss in accuracy.

In contrast, if we difference the first column we obtain the table:

x	$f(x)$	Δ	Δ^2	Δ^3
10	0000			
		414		
11	0414		-36	
		378		5
12	0792		-31	
		347		6
13	1139		-25	
		322		
14	1461			

In this case we should require a polynomial of degree about three for interpolation, judging from the higher differences; hence this tabular interval is inconveniently large for a working 4D log table.

3.2. Interpolation using divided differences: Newton's formula

Suppose that we have a table with constant tabular interval and we want to interpolate a value. The very act of putting in the 'new tabular point' at which we wish to determine f destroys the original uniformity of the tabular interval by producing two smaller intervals about the new point, of length θh and $(1-\theta)h$, so that we must for the present work in terms of divided differences.

The problem remains essentially the same if the tabular intervals are generally unequal as in §2.6, and we shall assume this for the time being. We now seek to represent the value $f(x)$ at a non-tabular value x in terms of finite differences for the tabular points, and this can be done in the following way. The original definitions for divided differences of successively higher order can be rewritten as

$$f(x) = f(x_0) + (x-x_0)[xx_0],$$
$$[xx_0] = [x_0x_1] + (x-x_1)[xx_0x_1],$$
$$[xx_0x_1] = [x_0x_1x_2] + (x-x_2)[xx_0x_1x_2],$$

. . . .

$$[xx_0x_1 \ldots x_{n-1}] = [x_0x_1 \ldots x_n] + (x-x_n)[xx_0x_1 \ldots x_n].$$

By successive substitution, for $[xx_0]$ in the first equation from the second, for $[xx_0x_1]$ in the second from the third, and so on, we obtain

$$f(x) = f(x_0) + (x-x_0)[x_0x_1] + (x-x_0)(x-x_1)[x_0x_1x_2] + \ldots$$
$$+ (x-x_0)(x-x_1) \ldots (x-x_{n-1})[x_0x_1x_2 \ldots x_n] + R_n(x)$$
$$= P_n(x) + R_n(x),$$

where

$$R_n(x) = (x-x_0)(x-x_1)(x-x_2) \ldots (x-x_n)[xx_0x_1 \ldots x_n].$$

It should be noted that $x_0, x_1, \ldots, x_n$ can be *any selected ordering* of the n given tabular points, and that x can lie anywhere in the general range of the tabular points (and in particular need not be greater than x_0).

This expression for $f(x)$ would immediately provide an interpolation formula of the kind we are seeking ($f(x)$ given in terms of tabular values and their divided differences) if only we could assert that the remainder term $R_n(x)$ was small relative to $f(x)$. When the tabulated function is a kth degree polynomial, divided differences beyond the kth are all zero and $R_n(x) \equiv 0$ provided that $n > k$: hence we can

INTERPOLATION

interpolate polynomials exactly (or, to be more precise, to within rounding error).

Example 5. Find $(9\cdot3)^3$ from the table of example 3.

We can identify $x_0, x_1, \ldots$ with the given tabular points in any order we please, and it will be sensible to select them in increasing order of distance from the non-tabular value $9\cdot3$ so that the factors $(x-x_i)$ in $P_n(x)$ may be kept as small as possible. Hence take $x_0=9$, $x_1=12$, $x_2=5$, $x_3=14$, $x_4=15$, and $x_5=0$; and then

$$f(9\cdot3) = 729 + 0\cdot3 \times 333 + 0\cdot3 \times (-2\cdot7) \times 26 + 0\cdot3 \times (-2\cdot7) \\ \times 4\cdot3 \times 1 + 0\cdot3 \times (-2\cdot7) \times 4\cdot3 \times (-4\cdot7) \times 0 + 0 \\ = 804\cdot357$$

This result is, of course, exact (and also trivial, though it serves very well as a demonstration). Normally we should work to a specified number of decimals and there would be rounding error in the table and in the final result.

The functions we have to interpolate will not usually be polynomials, but whenever they are well suited to polynomial approximation the higher divided differences will be small (and indeed this is how we judge suitability). In such cases we use the interpolation expression without remainder,

$$f(x) \simeq f(x_0) + (x-x_0)[x_0 x_1] + (x-x_0)(x-x_1)[x_0 x_1 x_2] + \\ \ldots + (x-x_0)(x-x_1) \ldots (x-x_{n-1})[x_0 x_1 \ldots x_n].$$

This is *Newton's interpolation formula*. The choice of order (n) for the approximation is a matter for the judgment of the computer and depends on the behaviour of the higher differences in the table.

The truncated Newton interpolation formula

$$f(x) \simeq P_n(x)$$

is a polynomial approximation of degree n in which the coefficients are given in terms of the divided differences of the table. The remainder $R_n(x)$ may be regarded as an error term; and provided that $f(x)$ has $n+1$ derivatives in

NEWTON'S FORMULA

a range including $x_0, x_1, \ldots, x_n$, it can also be expressed as

$$R_n(x) = (x-x_0)(x-x_1) \ldots (x-x_n) f^{(n+1)}(\xi)/(n+1)!,$$

for at least one value ξ in the range (see problem 3 below). It may be seen from this form for the remainder or error term that for a given function we shall have a small interpolation error if the tabular intervals are all short and if we refrain from introducing tabular values distant from the level x.

A considerable practical advantage of Newton's formula is that we can at any stage add or utilize an additional value anywhere in the range without disturbing the terms already calculated. In general $f(x) \neq P_n(x)$ except at the tabular points, but the error $R_n(x)$ lies within assignable limits; and for small R_n the tabular points $x_0, \ldots, x_n$ should be chosen as nearly as possible symmetric about x. $P_n(x)$ is the *smoothest* function that agrees with $f(x)$ at the tabular points, as $d^{n+1}P_n(x)/dx^{n+1} = 0$ for all x.

It may be worth emphasizing that the symbol $\eqsim$ is used in the expression $f(x) \eqsim P_n(x)$ to show that the right-hand side is a polynomial representation and that some question of judgment is involved in the selection of $P_n(x)$. We are asserting that the function is sufficiently well defined by the table of values and is well-represented by a polynomial of appropriate degree. Many authors use an ordinary equality sign, but we wish to draw attention to these steps of approximation.

Problem 3. Show that when f has n derivatives there is a relation of the form $[x_0 x_1 x_2 \ldots x_n] = f^{(n)}(\xi)/n!$.

Since $f(x) = P_n(x) + R_n(x)$ it follows that R_n has n derivatives too. Hence, using Rolle's theorem, as $R_n(x)$ vanishes at the $n+1$ points $x_0, x_1, \ldots x_n$ it follows that $R_n'(x)$ has at least n zeros in the range, $R_n''(x)$ at least $n-1$ zeros, ... and ultimately that $R_n^{(n)}(x)$ has at least one zero, at some intermediate point ξ of the range. But

$$f^{(n)}(x) = P_n^{(n)}(x) + R_n^{(n)}(x) = n! [x_0 x_1 \ldots x_n] + R_n^{(n)}(x);$$

and hence $[x_0 x_1 x_2 \ldots x_n] = f^{(n)}(\xi)/n!$.

INTERPOLATION

3.3. Lagrange interpolation formula

For some applications it is more convenient to have a matching interpolation polynomial with coefficients which are given directly in terms of the tabulated values of the function, as in the case of the *Lagrange interpolation formula*,

$$f(x) \simeq L_n(x)$$
$$= \frac{(x-x_1)(x-x_2)\ldots(x-x_n)}{(x_0-x_1)(x_0-x_2)\ldots(x_0-x_n)} f(x_0) +$$
$$+ \frac{(x-x_0)(x-x_2)\ldots(x-x_n)}{(x_1-x_0)(x_1-x_2)\ldots(x_1-x_n)} f(x_1) + \ldots$$
$$+ \frac{(x-x_0)(x-x_1)\ldots(x-x_{n-1})}{(x_n-x_0)(x_n-x_1)\ldots(x_n-x_{n-1})} f(x_n).$$

The polynomials L_n and P_n are identically equal. For L_n is a polynomial of degree n which takes the values $f(x_i)$ at the tabular points x_i ($i = 0, 1, \ldots, n$). Moreover, $L_n(x) - P_n(x)$ is a polynomial of degree n with the $n+1$ zeros x_i, and hence is identically zero;* thus L_n is merely another form of the Newton polynomial P_n.

In practice the Lagrange polynomial $L_n(x)$ is less convenient for interpolation than $P_n(x)$. The value $f(x)$ must in general be more closely related to *neighbouring* tabular values f_i than to distant ones; this special dependence on neighbouring values is recognized implicitly in the form of $P_n(x)$ and it is relatively easy to adjust the degree of the approximation to suit local conditions, while in $L_n(x)$ all arguments appear symmetrically so that all terms must be calculated and adjustments are less easy.

3.4. Interpolation in a table with equal intervals

Most interpolation will be in tables with *constant tabular interval* and we shall now develop from Newton's formula

---

* Recall that: just one polynomial of degree less than or equal to n can be fitted to $n+1$ distinct points; a polynomial of degree n which *vanishes* at $n+1$ distinct points is identically zero.

3.4 INTERPOLATION WITH EQUAL INTERVALS

other forms appropriate for use with constant interval tables. If we want the approximate value $f(x)$ it is natural to choose x_0 as the algebraically smaller adjacent tabular point to x, so that $x = x_0 + \theta h$, where $0 < \theta < 1$ and h is the tabular interval; beyond this the choice of arguments need not be the natural order, and indeed we can obtain a variety of interpolation formulae by taking different orderings of the tabular points.

x_0	x_0	f_0			
			Δf_0		
x_1	$x_0 + h$	f_1		$\Delta^2 f_0$	
			Δf_1		$\Delta^3 f_0$
x_2	$x_0 + 2h$	f_2		$\Delta^2 f_1$		$\Delta^4 f_0$
			Δf_2		$\Delta^3 f_1$
x_3	$x_0 + 3h$	f_3		$\Delta^2 f_2$	
			Δf_3		
x_4	$x_0 + 4h$	f_4			
.	.	. . . .			

The simplest case is that of interpolation near the head of a table, in which case we identify the points $x_0, x_1, \ldots$ of Newton's formula with the tabular points in increasing sequence (i.e. at progressively greater distance from x). We must use forward differences as these provide the only higher-order differences available near the head of a table, and hence $x_k = x_0 + kh$, $x - x_k = (\theta - k)h$. Bearing in mind that $[x_0 x_1 \ldots x_n] = \Delta^n f_0 / (n! \, h^n)$, we have by direct substitution in Newton's formula

$$f(x) = f_0 + \frac{\theta}{1!} \Delta f_0 + \frac{\theta(\theta - 1)}{2!} \Delta^2 f_0 + \frac{\theta(\theta - 1)(\theta - 2)}{3!} \Delta^3 f_0 +$$
$$\ldots + \frac{\theta(\theta - 1) \ldots (\theta - n)}{(n+1)!} h^{n+1} f^{(n+1)}(\xi).$$

This is the *Newton-Gregory forward difference interpolation formula* for use near the start of a table, mainly in the interval (x_0, x_1).

In a similar way we can construct a Newton-Gregory

INTERPOLATION

backward difference interpolation formula; however, we shall be more interested in a *Newton-Gregory backward difference extrapolation formula* which may be used for extending a table beyond its last value* $f(x_n)$. Thus, if $x = x_n + \theta h$, we obtain

$$f(x) \simeq f_n + \frac{\theta}{1!}\nabla f_n + \frac{\theta(1+\theta)}{2!}\nabla^2 f_n + \ldots + \frac{\theta(1+\theta)\ldots(n-1+\theta)}{n!}\nabla^n f_n.$$

Most interpolation will be in the interior of tables, where we use central differences so as to give equal weight to the tabular values on either side of the working level. In this case we can introduce tabular values roughly symmetrically about the interpolation level if we identify the $x_0, x_1, \ldots$ of Newton's formula with the tabular points in order of increasing distance from x starting with the point algebraically just below and alternating above and below, as

	$x_0 - 3h$	f_{-3}			
x_4	$x_0 - 2h$	f_{-2}			
			$\delta f_{-3/2}$		
x_2	$x_0 - h$	f_{-1}		$\delta^2 f_{-1}$	
			$\delta f_{-1/2}$		$\delta^3 f_{-1/2}$
x_0	x_0	f_0		$\delta^2 f_0$		$\delta^4 f_0$
			$\delta f_{1/2}$		$\delta^3 f_{1/2}$
x_1	$x_0 + h$	f_1		$\delta^2 f_1$		$\delta^4 f_1$
			$\delta f_{3/2}$		$\delta^3 f_{3/2}$
x_3	$x_0 + 2h$	f_2		$\delta^2 f_2$	
			$\delta f_{5/2}$		
x_5	$x_0 + 3h$	f_3			
.	$x_0 + 4h$	f_4			
.	.	.	.	.	.

illustrated. In this case $x = x_0 + \theta h$, $x - x_{2k} = (\theta + k)h$,

---

* Extrapolation is a potentially more dangerous pursuit than interpolation and we can relatively easily extrapolate functions into regions where they do not exist!

3.4 INTERPOLATION WITH EQUAL INTERVALS

$x - x_{2k-1} = (\theta - k)h$, and $[x_0 x_1 \ldots x_{2n}] = \delta^{2n} f_0 / (2n)! \, h^{2n}$, $[x_0 x_1 \ldots x_{2n+1}] = \delta^{2n+1} f_{1/2} / (2n+1)! \, h^{2n+1}$ on account of the selection order of the tabular points; and on substituting in Newton's formula

$$f(x) \simeq f_0 + (x - x_0)[x_0 x_1] + (x - x_0)(x - x_1)[x_0 x_1 x_2] + \ldots$$

we obtain the 'basic Newton formula' for constant interval interpolation

$$f(x) \simeq f_0 + \theta \delta f_{1/2} + \frac{\theta(\theta-1)}{2!} \delta^2 f_0 + \frac{\theta(\theta-1)(\theta+1)}{3!} \delta^3 f_{1/2} + \\ + \frac{\theta(\theta-1)(\theta+1)(\theta-2)}{4!} \delta f^4{}_0 + \ldots .$$

Note that if we sometimes write our formulae as unterminated series this does not mean that they are to be regarded in any sense as infinite series, but only that the truncation point is left to the discretion of the computer in each case; if polynomial representation is suitable, the error will at first decrease as we take extra terms, but beyond a certain stage it will generally increase again. The basic Newton formula can be rearranged in more useful forms. (1) For interpolation over the inner part of the interval $x_0 x_1$ we can derive a form symmetrical about the midpoint $x_0 + \frac{1}{2}h$ by rewriting even differences as, e.g.

$$\delta^2 f_0 = \tfrac{1}{2}(\delta^2 f_0 + \delta^2 f_1) + \tfrac{1}{2}(\delta^2 f_0 - \delta^2 f_1) = \tfrac{1}{2}(\delta^2 f_0 + \delta^2 f_1) - \tfrac{1}{2}\delta^3 f_{1/2}.$$

Thus

$$f(x) \simeq \{\tfrac{1}{2}(f_0 + f_1) + \tfrac{1}{2}(f_0 - f_1)\} + \theta \delta f_{1/2} + \\ + \frac{\theta(\theta-1)}{2!} \{\tfrac{1}{2}(\delta^2 f_0 + \delta^2 f_1) + \tfrac{1}{2}(\delta^2 f_0 - \delta^2 f_1)\} + \ldots$$

$$\simeq \tfrac{1}{2}(f_0 + f_1) + (\theta - \tfrac{1}{2})\delta f_{1/2} + \frac{\theta(\theta-1)}{2.2!} (\delta^2 f_0 + \delta^2 f_1) + \\ + \frac{\theta(\theta-\tfrac{1}{2})(\theta-1)}{3!} \delta^3 f_{1/2} + \\ + \frac{(\theta+1)\theta(\theta-1)(\theta-2)}{2.4!} (\delta^4 f_0 + \delta^4 f_1) + \ldots .$$

D

INTERPOLATION

This is *Bessel's interpolation formula*. Bessel's interpolation formula is sometimes written

$$f(x) \simeq \mu f_{1/2} + (\theta - \tfrac{1}{2})\delta f_{1/2} + \frac{\theta(\theta-1)}{2!}\mu\delta^2 f_{1/2} + \frac{\theta(\theta-\tfrac{1}{2})(\theta-1)}{3!}\delta^3 f_{1/2} + \ldots$$

where the *averaging symbol* μ denotes the mean of two adjacent elements of a vertical column; thus $2\mu f_{1/2} = f_0 + f_1$, etc. It is also written

$$f(x) \simeq f_0 + \theta\delta f_{1/2} + B''(\delta^2 f_0 + \delta^2 f_1) + B'''\delta^3 f_{1/2} + B^{iv}(\delta^4 f_0 + \delta^4 f_1) + \ldots,$$

where $B'' = \tfrac{1}{2}\theta(\theta-1)/2!$, $B''' = \theta(\theta-\tfrac{1}{2})(\theta-1)/3!$, ... are the Bessel coefficients.

(2) Some sets of tables include auxiliary sub-tables of *even-order* central differences (as these lie on tabular lines); to obtain an appropriate interpolation formula we need only replace the odd-order differences in the basic Newton formula by the one-lower-order even-difference combinations,

$$f(x) \simeq f_0 + \theta(f_1 - f_0) + \frac{\theta(\theta-1)}{2!}\delta^2 f_0 + \frac{\theta(\theta-1)(\theta+1)}{3!}(\delta^2 f_1 - \delta^2 f_0) + \ldots.$$

Taking $\phi = 1 - \theta$,

$$f(x) \simeq \phi\left\{f_0 + \frac{\phi^2 - 1^2}{3!}\delta^2 f_0 + \frac{(\phi^2 - 1^2)(\phi^2 - 2^2)}{5!}\delta^4 f_0 + \ldots\right\}$$
$$+ \theta\left\{f_1 + \frac{\theta^2 - 1^2}{3!}\delta^2 f_1 + \frac{(\theta^2 - 1^2)(\theta^2 - 2^2)}{5!}\delta^4 f_1 + \ldots\right\}.$$

This is *Everett's interpolation formula*.

(3) We can obtain a form symmetrical about tabular points as follows:

3.4 INTERPOLATION WITH EQUAL INTERVALS

$$f(x) \simeq f_0 + \theta(\delta f_{1/2} - \tfrac{1}{2}\delta^2 f_0) + \frac{1}{2!}\theta^2\delta^2 f_0 + \frac{1}{3!}\theta(\theta^2-1)(\delta^3 f_{1/2} - \tfrac{1}{2}\delta^4 f_0)$$
$$+ \frac{1}{4!}\theta^2(\theta^2-1)\delta^4 f_0 + \ldots$$
$$= f_0 + \tfrac{1}{2}\theta(\delta f_{-1/2} + \delta f_{1/2}) + \frac{1}{2!}\theta^2\delta^2 f_0 +$$
$$+ \frac{1}{2.3!}\theta(\theta^2-1)(\delta^3 f_{-1/2} + \delta^3 f_{1/2}) + \ldots$$
$$= f_0 + \theta\mu\delta f_0 + \frac{1}{2!}\theta^2\delta^2 f_0 + \frac{1}{3!}\theta(\theta^2-1)\mu\delta^3 f_0 +$$
$$+ \frac{1}{4!}\theta^2(\theta^2-1)\delta^4 f_0 + \ldots,$$

and this is *Stirling's interpolation formula*; we shall use this form later in deriving differentiation and integration processes.

A useful formula for *half-way interpolation* follows from the Bessel or Everett forms by taking $\theta = \tfrac{1}{2}$:

$$f_{1/2} \simeq \tfrac{1}{2}\Big\{(f_0+f_1) - \tfrac{1}{8}(\delta^2 f_0 + \delta^2 f_1) + \tfrac{3}{128}(\delta^4 f_0 + \delta^4 f_1) -$$
$$- \tfrac{5}{1024}(\delta^6 f_0 + \delta^6 f_1)\Big\} + O(h^8).$$

This is simple to apply and useful as a first step in breaking down a table with large interval. Note that this form of expression indicates approximation and also shows (in terms of the interval) the order of magnitude of the error term.

Example 6. Use Bessel's interpolation formula to find $f(10/3)$ from the table for $f(x) = 1/x$ given on page 19.

Here $\theta = (10/3 - 3\cdot3)/(3\cdot4 - 3\cdot3) = 1/3$, so that the coefficients of the Bessel formula are $\theta(\theta-1)/2.2! = -1/18$, $\theta(\theta-\tfrac{1}{2})(\theta-1)/3! = 1/162$, $(\theta+1)\theta(\theta-1)(\theta-2)/2.4! = 5/486$, ... Thus,

INTERPOLATION

$$f(\tfrac{10}{3}) \simeq \cdot 303030 + \tfrac{1}{3}(-\cdot 008912) - \tfrac{1}{18}(\cdot 000558 + \cdot 000508) +$$
$$+ \tfrac{1}{162}(-\cdot 000050) + \ldots$$
$$= \cdot 303030 - \cdot 0029707 - \cdot 0000592 - \cdot 0000003 + \ldots$$
$$= \cdot 300000_{-2},$$

i.e. $f(3\tfrac{1}{3}) \simeq \cdot 300000$ on rounding off.

Note: (i) that Bessel's formula up to first differences is just linear interpolation, and is most easily calculated as $f_0 + \theta \delta f_{1/2}$; (ii) the guard figure kept as a subscript to minimize rounding error (which might otherwise accumulate undesirably); (iii) that the contribution from $\delta^3 f$ is just worth retaining, that from $\delta^4 f$ is not.

3.5. The use of Bessel and Everett interpolation: throwback

For an isolated interpolation we shall use the formula with which we are most familiar, but for a run of interpolations it will be advantageous to select the 'best' formula.

Bessel's formula is one of the most generally useful. In practice it is often sufficient to work to second differences only (giving a first correction to linear interpolation), though sometimes third and occasionally higher differences have to be taken into account. We can make a working estimate of the critical size at which each order of differences becomes negligible in its effect on interpolation by finding the greatest magnitude of each Bessel coefficient for $0 < \theta < 1$; thus $\max|\theta(\theta-1)/2!| = \cdot 125$,
$$\max|\theta(\theta-\tfrac{1}{2})(\theta-1)/3!| = \cdot 00802,$$
$\max|(\theta+1)\theta(\theta-1)(\theta-2)/4!| = \cdot 02344$, and hence we can neglect $\delta^2 f \leqslant 4$, $\delta^3 f \leqslant 60$, $\delta^4 f \leqslant 20$ units in the last decimal with consequent error less than a $\tfrac{1}{2}$ unit in the last decimal digit. A better estimate of the error in any particular case can be found from the error term, but these rough values are sufficient for normal working.

Mathematical tables often include some finite differences.

3.5 BESSEL AND EVERETT INTERPOLATION

If only first differences are given the table will take linear interpolation; but when higher differences are needed it is common to print only $\delta^2 f$ and $\delta^4 f$ because of the difficulty involved in printing odd differences on half-lines, and in this case Everett's formula will obviously be more convenient. A further advantage is that second difference terms in Everett's formula take complete account of $\delta^3 f$, and similarly Everett up to fourth differences actually incorporates the effects of the first *five* differences; hence it will be superior when third differences cannot be neglected in Bessel's formula ($\delta^3 f > 60$) but fourth differences are negligible, or when fourth and higher differences are significant (as when many decimal figures are needed, or when the tabular interval is large). It can easily be shown that $\delta^2 f \leqslant 4$ and $\delta^3 f \leqslant 20$ can be neglected in applying Everett's formula with a consequent error of less than half a unit.

In an effective interpolation formula we hope that there will be few terms to calculate and yet we must avoid excessive truncation error, and we can make gains in both these directions by taking partial account of one of the higher differences in the procedure known as *throwback*. For illustration, in Bessel's formula the whole contribution from the second and fourth differences is

$$\frac{\theta(\theta-1)}{2.2!}(\delta^2 f_0 + \delta^2 f_1) + \frac{(\theta+1)\theta(\theta-1)(\theta-2)}{2.4!}(\delta^4 f_0 + \delta^4 f_1)$$
$$= \frac{\theta(\theta-1)}{2.2!}\left\{(\delta^2 f_0 + \delta^2 f_1) - \frac{1}{12}(1+\theta)(2-\theta)(\delta^4 f_0 + \delta^4 f_1)\right\}.$$

By a fortunate numerical chance the coefficient $(1+\theta)(2-\theta)/12$ varies little over the range $0 < \theta < 1$; indeed it lies between $\cdot 1800$ and $\cdot 1875$ for $\cdot 2 < \theta < \cdot 8$, and near the ends of the θ-range the whole contribution is small because of the outer factor $\theta(\theta-1)/4$. Hence we get a *very good approximation* to a Bessel formula carried to

INTERPOLATION

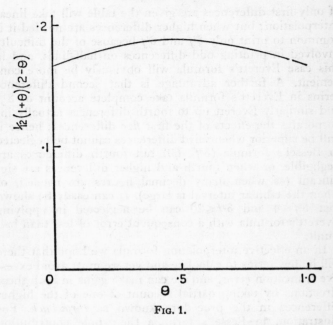

Fig. 1.

fourth differences if we work to third differences in the formula and use the *modified second differences*

$$\delta_m^2 f_i = \delta^2 f_i - C\delta^4 f_i,$$

where C is a constant, and may be taken as 0·184; this is known as throwing back the fourth difference on the second. Some accuracy is lost (though this throwback error is $<\frac{1}{2}$ unit for $\delta^4 f = 1{,}000$ units!), but we make a substantial saving in labour, and this is precisely the sort of net balance we always seek in designing numerical processes.

In the same way we can throw back fourth differences on the second in Everett's formula, and again we use $\delta_m^2 f_i = \delta^2 f_i - 0{\cdot}184\,\delta^4 f_i$ in place of $\delta^2 f_i$ and the modified Everett formula is

3.5 BESSEL AND EVERETT INTERPOLATION

$$f(x) \doteqdot (1-\theta)f_0 + \theta f_1 - \tfrac{1}{6}\theta(\theta-1)(\theta-2)\delta_m^2 f_0 + \\ + \tfrac{1}{6}(\theta+1)\theta(\theta-1)\delta_m^2 f_1.$$

Summarizing: (i) near the start or finish of a table use a Gregory formula; (ii) when $\delta^3 f < 60$ units in the last digit use Bessel up to second differences; (iii) when $\delta^3 f > 60$ but $\delta^4 f < 1{,}000$ units (or 400 units if $\delta^4 f_0$ and $\delta^4 f_1$ are of opposite sign) use Everett with throwback from fourth to second differences; (iv) when $\delta^4 f > 1{,}000$ units use throwback with higher differences.

Example 7. Find $f(2/3)$ from the table for $f = 1/x$ given below.

In the following table $h = \cdot 05$ and $x = 2/3 = \cdot 65 + h/3$, so that $\theta = 1/3$. Third differences cannot be neglected, but as fourth differences are less than a thousand we should normally use Everett with throwback, and we need then take no further account of $\delta^4 f$. As an illustration this particular example will be worked using: (i) Bessel to $\delta^4 f$; (ii) Bessel to $\delta^3 f$ with throwback; (iii) Everett to $\delta_m^2 f$.

x	$f(x)$	δ	δ^2	δ^3	δ^4	$-C\delta^4$	δ_m^2	$2\mu\delta_m^2$
·50	2·00000							
		−18182						
·55	1·81818		3031					
		−15151		−701				
·60	1·66667		2330		203	−37	2293	
		−12821		−498				
·65	1·53846		1832		131	−24	1808	
		−10989		−367				3256
·70	1·42857		1465		93	−17	1448	
		−9524		−274				
·75	1·33333		1191		63	−12	1179	
		−8333		−211				
·80	1·25000		980					
		−7353						
·85	1·17647							

INTERPOLATION

Hence:

(i)

$$\begin{aligned}
\theta(\theta-1)/4 &= -1/18 \\
\theta(\theta-\tfrac{1}{2})(\theta-1)/6 &= 1/162 \\
(\theta+1)\theta(\theta-1)(\theta-2)/48 &= 5/486 \\
f_0 &= 1\cdot 53846 \\
\theta\delta f_{1/2} &= -\cdot 03663_0 \\
-(\delta^2 f_0 + \delta^2 f_1)/18 &= -\cdot 00183_2 \\
\delta^3 f_{1/2}/162 &= -\cdot 00002_3 \\
5(\delta^4 f_0 + \delta^4 f_1)/486 &= \cdot 00002_3 \\
\hline
& 1\cdot 49999_8
\end{aligned}$$

(ii)

$$\begin{aligned}
f_0 &= 1\cdot 53846 \\
\delta f_{1/2}/3 &= -\cdot 03663_0 \\
-(\delta_m^2 f_0 + \delta_m^2 f_1)/18 &= -\cdot 00180_9 \\
\delta^3 f_{1/2}/162 &= -\cdot 00002_3 \\
\hline
& 1\cdot 49999_8
\end{aligned}$$

(iii)

$$\begin{aligned}
\phi(\phi^2-1)/6 &= -1/81 \\
\theta(\theta^2-1)/6 &= -4/81 \\
f_0 &= 1\cdot 53846 \\
\delta f_{1/2}/3 &= -\cdot 03663_0 \\
-\delta_m^2 f_0/81 &= -\cdot 00111_6 \\
-4\delta_m^2 f_1/81 &= -\cdot 00071_5 \\
\hline
& 1\cdot 49999_9
\end{aligned}$$

In each case the rounded solution is $1\cdot 50000$.

Note: (i) that the subscript m denotes a difference modified by throwback; (ii) that the effect of neglecting $\delta^4 f$ is about 2 in the last place, and some account must be taken of $\delta^4 f$ if the interpolate is to be as accurate as the tabular entries; (iii) that successive terms do not necessarily decrease monotonically in magnitude; (iv) that the probable round-off fluctuation of ± 6 in $\delta^4 f$ has negligible effect.

3.6. Some uses of interpolation

(i) *Subtabulation.* The choice of tabular interval in the construction a table is influenced by the uses anticipated for the table, but also by the economics of book production; hence we shall sometimes find that the printed tabular values are too widely spaced in regions of special interest. The process of 'filling in' between tabular values (known in this context as *pivotal values*) by systematic interpolation

3.6 SOME USES OF INTERPOLATION

is called *subtabulation*. For example, if the initial tabular interval h is to be broken down to $h/10$, it may be worth while applying the halfway interpolation formula followed by Everett using throwback as necessary. Subtabulation should be carried at least to an interval such that accurate interpolation is obtained using no higher than second differences; once a table can be linearly interpolated there is no gain from further subtabulation.

Note: (i) that if a table of a complicated function is to be built up by calculation from a formula and a fast computing machine is not available, the easiest procedure may be to calculate at large interval and subtabulate; (ii) that after halfway interpolation the differences in a table will be quite different and higher differences will be considerably reduced (as 2^{-n}), and that the usefulness of subtabulation arises from the fact that $\delta^n f = 0(h^n)$; (iii) that a table of irregularly spaced values can first be reduced to regular interval using Newton's divided difference formula, and then subtabulated in the normal way.

(ii) *Inverse interpolation.* In the process of interpolation we seek the value of $f(x)$ for a specified non-tabular value x; the inverse process in which we seek the value of x corresponding to a specified non-tabular value $f(x)$ is also important (e.g. in extracting roots of the equation $f(x) = 0$), and this is called *inverse interpolation*. It can be carried out by successively closer subtabulation in progressively smaller neighbourhoods of the specified value $f(x)$ until linear interpolation is possible (note that this approach allows full scope to the computer for an inspired guess at the neighbourhood and interval for starting).

Inverse interpolation can also be carried out by methods of successive approximation; this approach can be illustrated using Everett's formula. Everett's formula to second differences can be rearranged in the form

$$\theta \simeq \frac{1}{f_1 - f_0}\left\{f(x) - f_0 + \frac{\theta(\theta-1)}{6}[(\theta-2)\delta^2 f_0 - (\theta+1)\delta^2 f_1] + \ldots\right\}.$$

Provided that the function $f(x)$ is well defined by its table

INTERPOLATION

of values, the terms involving second and higher differences will normally decrease fairly rapidly, so that we can expect to get a modest first approximation θ_1 by ignoring the contributions of second and higher differences, and taking

$$\theta_1 = \frac{f(x)-f_0}{f_1-f_0} = \frac{f(x)-f_0}{\delta f_{1/2}}.$$

To improve this approximation we must bring in the term in second differences, and this involves θ; but as this term will provide only a relatively small correction to θ_1, we can get a satisfactory second approximation θ_2 from

$$\theta_2 = \{f(x)-f_0 + \tfrac{1}{6}\theta_1(\theta_1-1)[(\theta_1-2)\delta^2 f_0 - (\theta_1+1)\delta^2 f_1]\}/\delta f_{1/2}$$

where the error introduced by writing θ_1 in place of θ in the second-difference term is likely to be of the same order as the higher-difference terms which are neglected. The accuracy of the approximation can be improved by the successive introduction of terms in higher-order differences (though only to the degree permitted by the quality of the table). Note that the number of decimals which should be carried at each stage can be estimated by making the rounding error roughly comparable with the first term neglected (and it should be stressed that there is no gain from carrying too many figures in the early stages of the calculation).

Example 8. Find the value of x for which $\log_e x = 0.3$ from the following table.

x	$\log_e x$	δ	δ^2	δ^3	δ^4	δ^5	δ^6
1·2	·182322		−6970		−292		−57
		80042		1036		75	
1·3	·262364		−5934		−217		−16
		74108		819		59	
1·4	·336472		−5115		−158		−23
		68993		661		36	
1·5	·405465		−4454		−122		−3
		64539		539		33	
1·6	·470004		−3915		−89		−21
		60624		450		12	
1·7	·530628		−3465		−77		+10

3.6 SOME USES OF INTERPOLATION

Working from the table,
$$\theta_1 = \frac{0 \cdot 3 - f_0}{f_1 - f_0} = \frac{\cdot 3 - \cdot 262364}{\cdot 336472 - \cdot 262364} = \cdot 507853.$$

Our best choice is probably to proceed with the value $\theta_1 = 0 \cdot 5$, if we take due account of ease of calculation at the next step and also the likely error in this term! Then, using Everett up to modified second differences (we may just as well introduce at this stage the small but not negligible contribution from the fourth difference) we have for the second approximation,

$$\theta_2 = \{f(x) - f_0 + \tfrac{1}{6}\theta_1(\theta_1-1)[(\theta_1-2)\delta_m^2 f_0 - (\theta_1+1)\delta_m^2 f_1]\}/\delta f_{1/2}$$
$$= \left\{\cdot 3 - \cdot 262364 - \frac{1}{24}\left[-\frac{3}{2}(-\cdot 005894) - \frac{3}{2}(-\cdot 005086)\right]\right\} \div$$
$$\div \cdot 074108$$
$$= \cdot 498593.$$

A recalculation of θ_2 using the improved approximation $\theta \simeq \cdot 49859$ in place of $\theta \simeq \cdot 5$ will now serve both to reassure us that no substantial error has arisen from use of the crude value $\theta_1 = \cdot 5$ in calculating θ_2 (i.e. that the approximation is in fact self-consistent to the second stage), and to check θ_2 against blunders.

$$\theta_2' = \{\cdot 3 - \cdot 262364 + \tfrac{1}{6} \times \cdot 49859 \times (-\cdot 50141)[-1 \cdot 50141 \times$$
$$\times (-\cdot 005894) - 1 \cdot 49859 \times (-\cdot 005086)]\}/\cdot 074108$$
$$= \cdot 498593.$$

There will be no gain in accuracy by proceeding further, as the fourth differences (of about 200) have been sufficiently taken into account by throwback and the contribution to the numerator from the sixth differences is negligible. Hence

$$x \simeq 1 \cdot 3 + \cdot 49859 \times 0 \cdot 1 = 1 \cdot 349859.$$

Note: (i) that the irregular appearance of the sixth differences is sufficiently explained by the probable round-off fluctuation of ± 22; (ii) that by recalculating θ_2 we avoid carrying on an unnecessary error to the next stage, and in this case if we had used $\theta_1 = \cdot 508$ we would have found $\theta_2 = \cdot 49859$ and might have carried on $\cdot 49860$; (iii) that processes of successive approximation in which each improved estimate is based on the previous one are called *iterative processes*; (iv) that modified differences

INTERPOLATION

are useful in inverse as in direct interpolation; (v) that the accuracy of the process is good provided that $\delta f_{1/2}$ is not small, but that for small $\delta f_{1/2}$ the rounding errors can be disastrously amplified, for example near a stationary point of $f(x)$, and it is then usually best to subtabulate until linear interpolation is possible; (vi) that an *initial survey* of the problem is absolutely essential so that the computer may have some idea of the sort of pitfalls that must be avoided; (vii) that when higher differences are large it may be best to subtabulate at interval $\frac{1}{2}h$ as a first step; (viii) that on no account should the interpolation formula be treated as a quadratic (or cubic) to be solved for θ, as this is a laborious and inaccurate approach.

3.7. General notes on interpolation

(i) Tables of Bessel and Everett coefficients for $\theta = 0(\cdot 001)1$ can be found in *Interpolation and Allied Tables* (H.M.S.O.), in *Chambers's Six Figure Mathematical Tables*, Vol. 2, and elsewhere.

Critical tables of coefficients are specially useful, and a set of critical tables for Bessel coefficients is given in *Chambers's Shorter Six Figure Mathematical Tables*. Bessel coefficients vary slowly with θ, and it is possible in quite a short table to write down *all* the values taken by a coefficient, for example $B'' = \cdot 001, \cdot 002, \cdot 003, \ldots$, and to list beside them the critical values defining sub-intervals of θ which correspond to each value of B''. The leading terms in the critical table for B'' are:

θ	B''
·000	
	— ·000
·002	
	— ·001
·006	
	— ·002
·010	
	— ·003
·014	
·	·

These critical tables are easy to work from and are very convenient.

(ii) If instead of a table of values we start from a function

EXERCISES

which can be differentiated with reasonable ease it may be more effective to evaluate the function and its first few derivatives at pivotal points of the range, and then to sub-tabulate using Taylor series based in turn on each successive pivotal point. The terms in the Taylor series are in general smaller than corresponding terms in any finite difference interpolation formula.

(iii) Scientific literature contains many accounts of (so-called) *semi-empirical theories*, in which dimensional and other incomplete arguments are used to derive solution formulae containing several disposable constants, and these constants are to be evaluated from experimental results to complete the solution. But *any* given segment of curve can be matched to specified accuracy by a polynomial, which is fully determined by its coefficients. Put in another way, *any* curve segment (or set of experimental points) can be matched by a function containing a sufficient number of disposable constants, and the relatively simple curves which ought to be the product of well-conceived experiments can normally be matched over moderate ranges (and within experimental accuracy) by quite low degree polynomials. Hence we must treat semi-empirical theories with great reserve; for the link between theory and experiment has been reduced (at worst) to one of curve-matching and we have forfeited the direct relationship which alone justifies our construction of theories.

EXERCISES ON CHAPTER THREE

Again exercises may easily be constructed from any set of tables.

1. Subtabulate the following table to produce a table of $f(x)$ for $x = 1 \cdot 5(1/30)2 \cdot 4$.

x	$f(x)$	x	$f(x)$	x	$f(x)$
1·5	·40546	1·9	·64185	2·3	·83291
1·6	·47000	2·0	·69315	2·4	·87547
1·7	·53063	2·1	·74194		
1·8	·58779	2·2	·78846		

INTERPOLATION

2. State which of the two following tables is suitable for linear interpolation and estimate the subtabulation needed in the other to make it suitable.

	(i)	(ii)
	·176809	·176916
	·177840	·177896
	·178872	·178891
	·179904	·179904
	·180937	·180937
	·181970	·181993
	·183003	·183075
	·184037	·184186

3. Find $f(4\cdot 2)$ from the table.

x	$f(x)$	x	$f(x)$	x	$f(x)$
2·0	·500000	4·0	·250000	6·0	·166667
2·5	·400000	4·5	·222222	6·5	·153846
3·0	·333333	5·0	·200000		
3·5	·285714	5·5	·181818		

4. The following extract is taken from the table of natural tangents in Castle's *Five-figure logarithms and other tables*.

degrees	0′	6′	12′	18′	24′	30′	36′	42′	48′	54′
67	2·35585	36733	37891	39058	40235	41421	42618	43825	45043	46270
68	2·47509	48758	50018	51289	52571	53865	56170	56487	57815	59156

degrees	mean differences				
	1	2	3	4	5
67	199	397	596	795	994
68	mean differences cease to be sufficiently accurate				

Find whether linear interpolation does in fact provide 5D accuracy, and determine the errors which will be introduced if the mean differences are used in calculating the tangents of 67° 3′, 67° 27′ and 67° 57′.

5. Evaluate $\tan \sqrt{x}$ when $x = \cdot 075$ radian, using the table for $\tan \sqrt{x}$ given in problem 4.

6. Subtabulate the following table to produce a table of $f(x)$ for $x = 0(\cdot 1)2\cdot 5$.

x	$f(x)$	x	$f(x)$	x	$f(x)$
0	1·00000	1·8	3·10747	2·3	5·03722
0·9	1·43309	1·9	3·41773	2·4	5·55695
1·4	2·15090	2·1	4·14431		
1·6	2·57746	2·2	4·56791		

EXERCISES

7. Show that the remainder term for Bessel's interpolation formula can be written as

$$\frac{\theta(\theta^2-1^2)(\theta^2-2^2)\ldots(\theta^2-\overline{n-1}^2)(\theta-n)}{(2n)!} h^{2n} f^{(2n)}(\xi),$$

and find the corresponding remainder terms for the Everett and Stirling interpolation formulae.

CHAPTER FOUR

Differentiation and Integration (Quadrature)

The process of interpolation consists essentially in fitting a polynomial to neighbouring tabular values of the function, and then in substituting the interpolation polynomial for the known or unknown function which the table of values is supposed to represent. The error of approximation involved in using $P_n(x)$ in place of the actual function $f(x)$ can be kept within any prescribed limits over the whole of a (finite) x-range* if we use polynomials of sufficiently high degree, which in practice implies retaining a sufficient number of terms in an interpolation formula. The general character of this process is illustrated in figure 2 where the continuous curve, which represents an interpolation polynomial $P_3(x)$, oscillates about the dashed curve representing the true function $f(x)$ and intersects it at tabular values (marked by crosses). The lateral separation has been magnified so that the figure may be clearer. If we wish to form derivatives or integrals from a table of values we have no option but to base our numerical processes on an approximating polynomial $P_n(x)$ (or some other approximation); otherwise we cannot find the derivative, for example, even at a tabular point.

Although the difference between $P_n(x)$ and $f(x)$ will be small over the working range of the approximation and zero at the tabular points, the gradients $P'_n(x)$ and $f'(x)$ are likely to differ more markedly, particularly near tabular points. This effect can be illustrated very simply by the

* Thus the approximation is uniformly valid over the range.

DIFFERENTIATION AND INTEGRATION

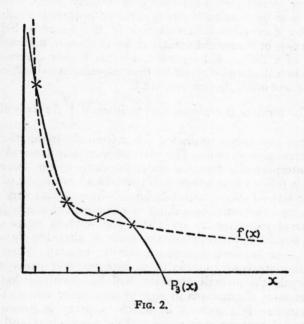

Fig. 2.

non-polynomial (and in fact quite useless) approximation $S(x) = f(x) + \epsilon \sin(2\pi x/h)$ to $f(x)$, where ϵ is small and h is the tabular interval. The greatest difference $|S(x) - f(x)|_{max}$ between $f(x)$ and its approximating function is ϵ at each halfway point. However the greatest difference between the gradients $|S'(x) - f'(x)|_{max} = |(2\pi\epsilon/h) \cos(2\pi x/h)|_{max}$ is $2\pi\epsilon/h$ at tabular points, and between kth derivatives $|S^{(k)}(x) - f^{(k)}(x)|_{max}$ is $(2\pi/h)^k \epsilon$ at tabular points for odd-order and halfway points for even-order derivatives. Thus for $h = \cdot 1$ the approximate first derivative is in error by about 60ϵ, and the kth derivative by about $(60)^k \epsilon$! Thus numerical differentiation is a process in which accuracy is *lost*, and in particular the determination of

DIFFERENTIATION AND INTEGRATION

higher derivatives numerically is usually unsatisfactory and is to be avoided if any alternative process is available (as by successive differentiation of the function before insertion of numerical values, or by successive differentiation of a differential equation, so that higher derivatives may be calculated directly by inserting numerical values of quite low order derivatives, etc.).

The numerical evaluation of an integral $\int_a^b f(x)dx$ is often termed *quadrature* to distinguish it from the integration of differential equations. The integral over any interval can be interpreted as the area under the curve in this interval, and it follows that integration involves an averaging of the error $f(x) - P_n(x)$ in each tabular sub-interval and very frequently substantial cancelling of error contributions from adjacent sub-intervals (as within the working range of a polynomial approximation the error is alternatively $+ve$ and $-ve$ between successive tabular points). When we accumulate contributions to an integral over many sub-intervals the overall accuracy will be greatly enhanced because of alternations in sign of local error terms. Thus quadrature is a process in which accuracy is *gained* (in powerful contrast with numerical differentiation), and this effect is sufficiently marked in practice for us to adopt a basic integration process that depends on quadratic (i.e. parabola) matching.

4.1. Numerical differentiation

Formulae for numerical differentiation can be obtained easily by differentiating the interpolation polynomial, and this can be carried out conveniently using one of our interpolation formulae. Derivatives are needed most commonly at a tabular point, and hence from §3.4 we select Stirling's interpolation formula as the form symmetric about tabular points. When this is differentiated,

NUMERICAL DIFFERENTIATION

$$f'(x) = \frac{df(x)}{dx} = \frac{1}{h}\frac{df}{d\theta} \simeq \frac{1}{h}\left\{\mu\delta f_0 + \theta\delta^2 f_0 + \frac{3\theta^2-1}{3!}\mu\delta^3 f_0 + \right.$$
$$\left. + \frac{2\theta(2\theta^2-1)}{4!}\delta^4 f_0 + \ldots \right\},$$

$$f''(x) = \frac{1}{h^2}\frac{d^2 f}{d\theta^2} \simeq \frac{1}{h^2}\left\{\delta^2 f_0 + \frac{6\theta}{3!}\mu\delta^3 f_0 + \frac{2(6\theta^2-1)}{4!}\delta^4 f_0 + \right.$$
$$\left. + \frac{10\theta(2\theta^2-3)}{5!}\mu\delta^5 f_0 + \ldots \right\},$$

and so on for higher derivatives, where $x = x_0 + \theta h$. These forms can be used directly to obtain derivatives at non-tabular points by suitable choice of θ; and with $\theta = 0$ we obtain *finite difference formulae for derivatives at tabular points* as,

$$hf_0' \simeq \mu\delta f_0 - \frac{1}{6}\mu\delta^3 f_0 + \frac{1}{30}\mu\delta^5 f_0 - \frac{1}{140}\mu\delta^7 f_0 + \ldots,$$

$$h^2 f_0'' \simeq \delta^2 f_0 - \frac{1}{12}\delta^4 f_0 + \frac{1}{90}\delta^6 f_0 - \frac{1}{560}\delta^8 f_0 + \ldots.$$

It must be emphasized again that these are not infinite series, but that the computer must select the truncation point appropriate to each application. Thus if a polynomial representation of degree five is appropriate for some application we use the form

$$f_0' \simeq \frac{1}{h}\left\{\mu\delta f_0 - \frac{1}{6}\mu\delta^3 f_0 + \frac{1}{30}\mu\delta^5 f_0\right\},$$

which can also be written in the form:*

$$f_0' \simeq \frac{1}{h}\left\{\mu\delta f_0 - \frac{1}{6}\mu\delta^3 f_0 + \frac{1}{30}\mu\delta^5 f_0\right\} + O(h^6).$$

The advantage of the latter form is that it gives an im-

---

* This form implies that in the case of a function suited to polynomial representation the first term neglected (i.e. $\mu\delta^7 f_0/140h$ here) is of order h^6, and gives a good estimate of the total error for small h.

DIFFERENTIATION AND INTEGRATION

mediate indication of the effect of a change in working interval on the truncation error.

It should be noted that these differentiation formulae are of the general form $f^{(k)} \simeq$ (*some combination of differences*)$/h^k$. The rounding errors in the (*combination of differences*) depend very little on the interval h, so that the consequent error in the derivative is proportional to h^{-k}, and will be greatly amplified even for low-order derivatives when h is small. Hence in differentiation we should *use as large an interval as is consistent with a proper representation of the function*, even though this may involve the introduction of a number of higher-order difference terms.

Example 9. Find $d(\cos x)/dx$ at $x = 0.5$ radians from a 6D table of $\cos x$ with interval 0.001.

From our experience with interpolation we might choose to work with as small a tabular interval as possible, ·001 in this case. Extracting that part of the difference table actually needed,

x	$f(x)$	δf	$\delta^2 f$
·499	·878062		−1
		−479	
·500	·877583		−1
		−480	
·501	·877103		−1.

The general formula for the first derivative is

$$hf'_0 \simeq \mu\delta f_0 - \frac{1}{6}\mu\delta^3 f_0 + \frac{1}{30}\mu\delta^5 f_0 - \ldots ,$$

and as third differences are negligible we shall truncate it after the first term, so that on substitution from the table $\cdot 001\, f'(\cdot 5) \simeq -\cdot 000479_5$, and $f'(\cdot 5) \simeq -\cdot 480$. Note that the maximum error in $\mu\delta f$ from rounding of the tabular values $f(x)$ is half a unit in the sixth decimal, and that in consequence $f'(\cdot 5)$ has a possible error of half a unit in the *third decimal place*!

Repeating the calculation with tabular entries at interval ·01 selected from the table,

4.1 NUMERICAL DIFFERENTIATION

·49	·886995		−88
		−4750	
·50	·877583		−88
		−4838	
·51	·872745		−88,

we again use the formula $hf'_0 \simeq \mu\delta f_0$ and hence $·01f'(·5) \simeq -·004794$, and $f'(·5) \simeq -·4794$ with a possible round-off error of half a unit in the fourth decimal digit. Again with $h=0·1$,

·4	·921061		−9203		93
		−43478		434	−5
·5	·877583		−8769		88
		−52247		522	−4
·6	·825336		−8247		84

$·1f'(·5) \simeq -·047862_5 - ·000079_7 - 000000_2$, and $f'(·5) \simeq -·47942$ with a possible error approaching one unit in the fifth decimal (arising from the round-off error in the two terms of the formula $hf'_0 \simeq \mu\delta f_0 - \tfrac{1}{6}\mu\delta^3 f_0$). Finally with $h=0·3$,

·2	·980067		−87547		7822	
		−102484		9155		−820
·5	·877583		−78392		7002	−622
		−180876		16157		−1442
·8	·696707		−62235		5560	

$·3f'(·5) \simeq -·141680 - ·002109_3 - ·000037_3 + \ldots$. Still further differences must be taken into account (i.e. a higher degree matching is needed), and the difficulty is that at the term $\dfrac{1}{30}\mu\delta^5 f_0$ the contribution of the *probable* rounding error is about 4×10^{-7} and is increasing more rapidly than the coefficients of the series decrease; hence we have reached a stage at which the rate of decrease of higher-order differences is too slow for numerical differentiation (i.e. the tabular values utilized do not define the function sufficiently well for this purpose).

Note: (i) that the coefficients of the differentiation series decrease rather slowly; (ii) that there are two distinct disadvantages of numerical differentiation, the amplification of round-off when the tabular interval is small and the successively poorer matching of the derivatives $P_n^{(k)}(x)$ and $f^{(k)}(x)$ as

DIFFERENTIATION AND INTEGRATION

the order k increases; (iii) that the calculation of $f''(\cdot 5)$ from the above data goes much worse, $(\cdot 001)^2 f''(\cdot 5) \simeq -\cdot 000001$, $(\cdot 01)^2 f''(\cdot 5) \simeq -\cdot 000088$, etc.

4.2. The integration of a specified integrand; Simpson's rule

The process of quadrature, that is the numerical evaluation of the definite integral $\int_a^b f(x)dx$ of some specified function $f(x)$, is of considerable importance, as we are often unable to find effective analytical processes for evaluating integrals even though the functional form of the integrand is known. The obvious way out of this difficulty is to approximate to the known integrand $f(x)$ with some function we *can* integrate, and we are immediately on familiar

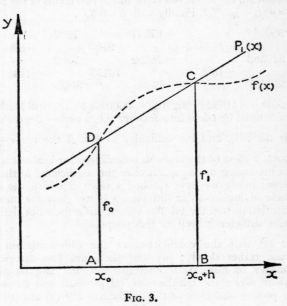

Fig. 3.

SIMPSON'S RULE

ground as it was for a very similar kind of approximation (to the supposed function underlying a set of tabular values) that we have introduced interpolation polynomials, and we are really rather good at integrating polynomials!

The simplest of interpolation polynomials is the chord joining two points of the curve $y = f(x)$,

$$y = P_1(x) \equiv f_0 + \theta \Delta f_0.$$

Hence we have the crude approximation

$$\int_{x_0}^{x_0+h} f(x)dx \simeq \int_{x_0}^{x_0+h} P_1(x)dx = h\int_0^1 (f_0 + \theta \Delta f_0)d\theta,$$

or

$$\int_{x_0}^{x_0+h} f(x)dx \simeq \tfrac{1}{2}h(f_0 + f_1),$$

which is known as the *trapezoidal rule*, and consists in replacing the integral by the area of the trapezium ABCD, as shown in figure 3. A substantial interval $a \leqslant x \leqslant b$ can be subdivided into a number of sub-intervals of length h in each of which the simple trapezoidal rule is applied, to give

$$\int_{a=x_0}^{b=x_0+nh} f(x)dx \simeq h(\tfrac{1}{2}f_0 + f_1 + f_2 + \ldots + f_{n-1} + \tfrac{1}{2}f_n).$$

The *trapezoidal rule* will obviously give a poor approximation unless the sub-interval h is short enough to identify each time an almost linear region of the curve $y = f(x)$.

A much better approximation can be obtained by fitting an arc of a parabola to the curve $y = f(x)$ at the three tabular points x_0, $x_0 + h$, $x_0 + 2h$; that is by using the interpolation polynomial $P_2(x)$ in an interval of length $2h$. Then

$$\int_{x_0}^{x_0+2h} f(x)dx$$

$$\simeq \int_{x_0}^{x_0+2h} P_2(x)dx = h\int_0^2 \left\{f_0 + \theta \Delta f_0 + \frac{\theta(\theta-1)}{2}\Delta^2 f_0\right\}d\theta,$$

or
$$\int_{x_0}^{x_0+2h} f(x)dx \simeq \frac{1}{3}h(f_0 + 4f_1 + f_2).$$

This is *Simpson's Rule*, and it provides a remarkably simple and effective process for numerical integration. We shall show that the direct application of Simpson's rule will in many cases provide sufficient accuracy, provided that the working interval h is sufficiently small; this will usually involve subdivision of the whole interval (a, b) of integration into an even number of sub-intervals, when

$$\int_{x_0}^{x_0+2nh} f(x)dx \simeq \frac{1}{3}h(f_0 + 4f_1 + 2f_2 + 4f_3 + 2f_4 + \ldots + 4f_{2n-1} + f_{2n}).$$

A variety of quadrature formulae can be obtained by using interpolation polynomials of successively higher degree with matching at a successively larger number of tabular points, or by suitable combination of other formulae in such a way as partially to cancel errors. However, these tend to be very largely of theoretical interest, as they involve subdivision of the range into awkward numbers of intervals (which are integer multiples of the degree n of the matching polynomial $P_n(x)$), and also they tend to involve a variety of coefficients so that they are less convenient in use and more liable to blunders. For example, the *three-eighths rule* is obtained using the matching polynomial $P_3(x)$ fitted at x_0, x_0+h, x_0+2h, x_0+3h; and Weddle's rule using $P_6(x)$ fitted at $x_0, \ldots, x_0+6h$ (see exercise 3 for these formulae).

We must have an estimate of the *error* involved in these quadrature processes before we can apply them confidently. Suppose that $f(x)$ is continuous in the range (x_0-h, x_0+h), and in this range has the Taylor expansion

$$f(x) = f_0 + xf_0' + \frac{1}{2!}x^2 f_0'' + \frac{1}{3!}x^3 f''' + \ldots + R_{2k},$$

where $R_{2k} = x^{2k} f^{(2k)}(\xi)/(2k)!$ for some $\xi = \xi(x)$ between

4.2　SIMPSON'S RULE

x_0 and x. Then on integrating over the symmetrically disposed range (x_0-h, x_0+h),

$$\int_{x_0-h}^{x_0+h} f(x)dx = 2h\left\{f_0 + \frac{1}{3}\frac{h^2}{2!}f_o'' + \frac{1}{5}\frac{h^4}{4!}f_0^{(4)} + \ldots + \frac{1}{2h}\int_{x_0-h}^{x_0+h} R_{2k}dx\right\}.$$

Simpson's rule can be expressed over the same range in terms of the derivatives $f_o^{(i)}$ by expanding $f_1 \equiv f(x_0+h)$ and $f_{-1} \equiv f(x_0-h)$ separately in Taylor series and combining the two terminated series,

$$\int_{x_0-h}^{x_0+h} f(x)dx \simeq \frac{1}{3}h(f_{-1} + 4f_0 + f_1)$$

$$= \frac{1}{3}h\left\{f_0 - hf_o' + \frac{1}{2!}h^2 f_o'' - \ldots + \frac{1}{(2k)!}h^{2k}f^{(2k)}(\xi_1)\right.$$
$$\left. + 4f_0 + f_0 + hf_o' + \frac{1}{2!}h^2 f_o'' + \ldots + \frac{1}{(2k)!}h^{2k}f^{(2k)}(\xi_2)\right\}$$
$$= 2h\left\{f_0 + \frac{1}{3}\frac{h^2}{2!}f_o'' + \frac{1}{3}\frac{h^4}{4!}f_0^{(4)} + \ldots\right\},$$

where $x_0-h \leqslant \xi_1 \leqslant x_0$ and $x_0 \leqslant \xi_2 \leqslant x_0+h$. By comparing these two forms it may be seen that the *correction* needed to Simpson's rule to give correct quadrature is

$$-\frac{1}{90}h^5 f_0^{(4)} - \frac{1}{1890}h^7 f_0^{(6)} - \ldots + \int_{x_0-h}^{x_0+h} R_{2k}dx -$$
$$-\frac{1}{3(2k)!}h^{2k+1}\{f^{(2k)}(\xi_1) + f^{(2k)}(\xi_2)\}.$$

If $f^{(2k)}(x)$ is bounded in (x_0-h, x_0+h) it is easy to show that the remainder term in this correction has the value $-\frac{4}{3}(k-1)h^{2k+1}f^{(2k)}(\xi)/(2k+1)!$ for some ξ lying in the range (x_0-h, x_0+h) but not necessarily equal to any of the

DIFFERENTIATION AND INTEGRATION

previous values. Provided that the interval h is chosen so that $f(x)$ is well-represented by a table of values at interval h, the first of these terms will dominate the correction and yet will itself be small. Thus we can integrate a very wide range of functions with Simpson's rule provided only that we are prepared to use a suitably chosen working interval, and in most cases this will involve subdivision of the range of integration; the extent of this subdivision will be a matter for judgment based on the character of the integrand. We shall also define the *error* $= -(correction)$.

The accuracy of formulae based on higher degree interpolation polynomials can be estimated in a similar way (see exercise 3). Formulae based on polynomials of even degree are found in general to be more effective than those of odd degree; thus the corrections expressed as single terms, are $-h^3 f_o''(\xi)/12$ for the trapezoidal rule, $-h^5 f_0^{(4)}(\xi)/90$ for Simpson's rule, and $-3h^5 f_0^{(4)}(\xi)/80$ for the three-eighths rule (where the ranges of integration have been taken as h, $2h$, and $3h$, respectively, and in each case ξ is some point within the integration range). Note that the accuracy of the three-eighths rule is slightly inferior to that of Simpson's rule when the same tabular values are used; it is, however, useful to complete an integration over an *odd* number of sub-intervals, when the three-eighths rule can be used for either the first three or the last three intervals and Simpson's rule for the remainder of the range.

The above integration formulae use tabular values lying within and at the ends of the range of integration, and they are sometimes known as *closed-type integration formulae*. We can also develop *open-type integration formulae*, which involve only tabular values within the integration range. For example, if we use a parabola fitted to the tabular values f_1, f_2, and f_3 to derive an integration formula over the more extensive range $(x_0, x_0 + 4h)$ we have*

---

* Note that $\theta = 0$ in the interpolation quadratic corresponds with the tabular point $x_1 \equiv x_0 + h$.

4.3 QUADRATURE FROM A TABLE OF VALUES

$$\int_{x_0}^{x_0+4h} f(x)dx \simeq h\int_{-1}^{3}\{f_1 + \theta\Delta f_1 + \tfrac{1}{2}\theta(\theta-1)\Delta^2 f_1\}d\theta$$
$$= \frac{4}{3}h(2f_1 - f_2 + 2f_3).$$

The correction term is $+14h^5 f^{(4)}(\xi)/45$. Closed-type integration formulae will obviously be more accurate than open-type formulae over the same range, as they make use of more information; hence closed-type formulae are usually used for quadrature, but open-type formulae are useful in solving differential equations.

4.3. Quadrature from a table of values

Formulae for quadrature from a table of finite differences can be obtained by integrating the interpolation formulae. Thus from Everett's formula we have

$$\int_{x_0}^{x_0+h} f(x)dx \simeq \frac{1}{2}h\Big\{(f_0 + f_1) - \frac{1}{12}(\delta^2 f_0 + \delta^2 f_1) +$$
$$+ \frac{11}{720}(\delta^4 f_0 + \delta^4 f_1) - \frac{191}{60480}(\delta^6 f_0 + \delta^6 f_1)\Big\} + O(h^9).$$

For integration over an extensive range we can add formulae of this type to span the range (using the results $\delta^2 f_0 + \delta^2 f_1 = \delta f_{\frac{3}{2}} - \delta f_{-\frac{1}{2}}$, etc.),

$$\int_{x_0}^{x_0+nh} f(x)dx \simeq h\Big\{\frac{1}{2}f_0 + f_1 + f_2 + \ldots + f_{n-1} + \frac{1}{2}f_n -$$
$$- \frac{1}{12}(\mu\delta f_n - \mu\delta f_0) + \frac{11}{720}(\mu\delta^3 f_n - \mu\delta^3 f_0) -$$
$$- \frac{191}{60480}(\mu\delta^5 f_n - \mu\delta^5 f_0) + \frac{2497}{3628800}(\mu\delta^7 f_n - \mu\delta^7 f_0)\Big\} + O(h^{10}).$$

These formulae represent improved forms of the trapezoidal rule with corrections which involve the differences near ends of the range of integration; they may also be regarded

DIFFERENTIATION AND INTEGRATION

as giving the error of the trapezoidal rule in terms of finite differences of the integrand.

Similarly, from integration of Stirling's formula,

$$\int_{x_0}^{x_0+2h} f(x)dx \simeq 2h\left\{f_1 + \frac{1}{6}\delta^2 f_1 - \frac{1}{180}\delta^4 f_1 + \frac{1}{1512}\delta^6 f_1\right\} + O(h^9)$$

$$= h\left\{\frac{1}{3}(f_0 + 4f_1 + f_2) - \frac{1}{90}\delta^4 f_1 + \frac{1}{756}\delta^6 f_1\right\} + O(h^9),$$

giving the corrected form of Simpson's rule in finite differences.

The choice of truncation point in each of the above formulae must be made separately in the context of each new problem, but the mode of writing the formulae adopted here indicates the dependence on working interval of the remainder in case of truncation after sixth or seventh differences. In practice it may often be sufficient to work with only two or three terms. We often omit remainder terms when writing formulae; for in fact we seldom bother to evaluate them but rather infer their magnitudes from the size and sequential behaviour of the terms which we do take into account. Thus we judge directly from the table of differences where a formula can be truncated without serious truncation error, and in so doing we select the degree of the approximating polynomial implicitly.

The formulae of this section (used in truncated form with one or more difference terms) take into account some tabular values *outside* the range of integration as well as those within and so correspond with the use of higher degree matching polynomials; they are known as *partial-range integration formulae*, and for a specified working interval they will be considerably more accurate provided that $f(x)$ is well-represented (and this can be judged from the size of higher differences).

Example 10. Evaluate $\int_0^5 e^x dx$ from the table

4.3 QUADRATURE FROM A TABLE OF VALUES

x	0	·5	1·0	1·5	2·0	2·5
e^x	1	1·6487	2·7183	4·4817	7·3891	12·1825
x	3·0		3·5	4·0	4·5	5·0
e^x	20·0855		33·1154	54·5982	90·0171	148·4132

The integrand e^x is a smooth function and should be easy to approximate except perhaps at the upper end of the range, where it is increasing more rapidly. The correct value of the integral to 4D is $e^5 - 1 = 147\cdot4132$. The trapezoidal rule with interval $h = 0\cdot5$ gives

$$\int_0^5 e^x dx \simeq \cdot5(\tfrac{1}{2} \times 1 + 1\cdot6487 + 2\cdot7183 + \ldots + 90\cdot0171 +$$
$$+ \tfrac{1}{2} \times 148\cdot4132)$$

$= 150\cdot4716$, with error $3\cdot0584$. Simpson's rule with the same working interval gives $\int_0^5 e^x dx \simeq (\cdot5/3)(1 + 6\cdot5948 + 5\cdot4366 + 17\cdot9268 + 14\cdot7782 + 48\cdot7300 + 40\cdot1710 + 132\cdot4616 + 109\cdot1964 + 360\cdot0684 + 148\cdot4132) = 147\cdot4628$, with error $\cdot0496$. The corrected trapezoidal rule needs finite differences calculated near the ends of the range, given in the following partial table:

— ·5	·6065		·1549		·0392		·0113	
		·3935		·1003		·0262		·0043
·0	1·0000		·2552		·0654		·0156	
		·6487		·1657		·0418		·0127
·5	1·6487		·4209		·1072		·0283	

. . . .

4·5	90·0171		22·9772		5·8643		1·4995	
		58·3961		14·9054		3·8059		·9669
5·0	148·4132		37·8826		9·6702		2·4664	
		96·2787		24·5756		6·2723		1·6022
5·5	244·6919		62·4582		15·9425		4·0686	

Hence $\int_0^5 e^x dx \simeq 150\cdot4716 - (\cdot5/12)(77\cdot3374 - \cdot5211) +$
$+ (\cdot5 \times 11/720)(19\cdot7405 - \cdot1330) - (\cdot5 \times 191/60480) \times$
$\times (5\cdot0391 - \cdot0340) + (\cdot5 \times 2497/3628800)(1\cdot2846 - \cdot0085)$
$= 147\cdot4132$.

Note: (i) that the simple trapezoidal rule is most seriously in error near the upper end of the range where e^x varies rapidly

DIFFERENTIATION AND INTEGRATION

with x and there is a contribution $1 \cdot 9464$ from $\int_4^5 e^x dx$ to the whole error $3 \cdot 0584$; (ii) that even with so large a sub-interval as $h = 0 \cdot 5$ the parabola fitting of Simpson's rule gives very reasonable accuracy when the function is smooth in this sub-interval; (iii) that a standard method of checking an integral evaluation is to repeat the calculation using a smaller interval (say half of that used initially) and when the two results agree to the accuracy required we can be confident both that the original interval was suitable and that the calculation is free from blunders; (iv) that when greater accuracy is needed the finite difference formula should be used; (v) that when the behaviour of the integrand varies appreciably from part to part of the integration range the most effective procedure may be to use working intervals of different length in different parts of the range so that the local accuracy of working is roughly uniform; (vi) that blunders are specially likely near points at which the working interval is changed, because of the interruption of a smooth routine, and as a safeguard the two calculations should be overlapped to give a check.

Problem 4. Evaluate $\int_0^1 \tan \sqrt{x} \, dx$ to four decimals.

Our first step might be to tabulate the integrand and its differences at interval $h = 0 \cdot 1$ (see next page). If we evaluate the integral by a normal application of Simpson's rule with working interval $h = 0 \cdot 1$ we have from the table,

$$\int_0^1 \tan \sqrt{x}\, dx \simeq \frac{\cdot 1}{3}(0 + 4 \times \cdot 3272 + 2 \times \cdot 4796 + \ldots + 4 \times 1 \cdot 3945 + \\ + 1 \cdot 5574) = 0 \cdot 8536.$$

This seems straightforward enough, and as a check on our working we repeat the calculation with $h = 0 \cdot 05$ and we find the value $0 \cdot 8553$. This poor agreement is not due to a blunder. It can be seen that differences near the top of the table are large even at high order, and this is related to the fact that the derivatives of $\tan \sqrt{x}$ are all infinite at $x = 0$. In fact all integrands which behave like $x^{1/2}$ or $x^{-1/2}$ near a terminal point $x = 0$ of an integration range (or like $|x-a|^{\pm \frac{1}{2}}$ near $x = a$) are troublesome

4.3 QUADRATURE FROM A TABLE OF VALUES

x	$\sqrt{x}$	$\tan\sqrt{x}$	δ	δ^2	δ^3	δ^4
0	0	0				
			3272			
·1	·3162	·3272		−1748		
			1524		1528	
·2	·4472	·4796		−220		−1383
			1304		145	
·3	·5477	·6100		−75		−83
			1229		62	
·4	·6325	·7329		−13		−24
			1216		38	
·5	·7071	·8545		+25		−8
			1241		30	
·6	·7746	·9786		55		−5
			1296		25	
·7	·8367	1·1082		80		+6
			1376		31	
·8	·8944	1·2458		111		0
			1487		31	
·9	·9487	1·3945		142		
			1629			
1·0	1·0000	1·5574				

even though their integrals converge. This kind of difficulty can be sidestepped in various ways.

(a) We can use an open-range formula for one set of intervals adjacent to $x=0$ (allowing us to 'stand back' a little from the trouble level $x=0$) and Simpson's rule in the remainder of the range. Thus with $h=0·1$,

$$\int_0^1 \tan\sqrt{x}\,dx = \int_0^{·4}\tan\sqrt{x}\,dx + \int_{·4}^1 \tan\sqrt{x}\,dx$$
$$\simeq (4\times ·1\times \tfrac{1}{3})(2\times ·3272 - ·4796 + 2\times ·6100) +$$
$$+ (·1\times \tfrac{1}{3})(·7329 + 4\times ·8545 + 2\times ·9786 + \ldots +$$
$$+ 1·5574)$$
$$= 0·8582.$$

The correction terms $-f^{(4)}(\xi)\times 10^{-5}/90$ for each application of Simpson's rule can be seen as negligibly small, but there is doubt about the term $+14f^{(4)}(\xi)\times 10^{-5}/45$ for the initial open-type formula. Probably the simplest way of resolving this doubt,

DIFFERENTIATION AND INTEGRATION

and at the same time checking the calculation, is again to repeat the process with working interval 0·05, and the 'improved' result is 0·8569. This decrease in working interval has clearly decreased the error from the open-range part of the integration but we cannot be sure that it has eliminated it. All that we can reasonably assert at this stage is that the integral lies between 0·8553 and 0·8569, as the correction terms are of opposite sign.

(b) In the neighbourhood of $x=0$ the integrand behaves as $\sqrt{x}$, and hence if we compute the integral of $(\tan \sqrt{x} - \sqrt{x})$ and subsequently use the result $\int_0^1 \sqrt{x}\, dx = 2/3$, the numerical work should be a good deal easier. Using values from the table above, we have from Simpson's rule with $h=0·1$,

$$\int_0^1 (\tan \sqrt{x} - \sqrt{x})dx \simeq ·1895, \text{ so that } \int_0^1 \tan \sqrt{x}\, dx \simeq ·8562.$$

When the integral is re-evaluated using working interval $h=0·05$ the result is again 0·8562, and this may be taken as correct to 4D.

(c) An important technique for dealing with awkward integrals is to convert them into differential equations in which the dependent variable is free from the singularity that has been causing the trouble. If we take $\int_0^x \tan \sqrt{u}\, du = g(x)y(x)$, where $y(x)$ is a function to be determined and $g(x)$ is selected as having a singularity related to that of the integrand, we obtain a differential equation on differentiating. With $g(x) = \sqrt{x}$ (same singularity as $\tan \sqrt{x}$ at $x=0$) we find $2xy' + y = 2\sqrt{x} \tan \sqrt{x}$; with $g(x) = x^{3/2}$ (when dg/dx has same singularity as integrand) we have $2xy' + 3y = 2\tan \sqrt{x}/\sqrt{x}$. For the solution of these differential equations see *Numerical Solution of Equations*.

(d) A change of variable often serves to reduce a difficult integral to more tractable form. In this case, if we take $\sqrt{x} = u$,

$$\int_0^1 \tan \sqrt{x}\, dx = 2\int_0^1 u \tan u\, du,$$

and numerical evaluation is considerably easier.

Note: (i) that for integration it is normally sufficient to work with four decimals to produce a result correct to 4D; (ii) that it is worth tabulating the differences of $\tan \sqrt{x}$ as a

4.3 QUADRATURE FROM A TABLE OF VALUES

check on the calculated tabular values; (iii) that care is needed in working with differences where there is a sign change that might be overlooked; (iv) that the variations in fourth differences are not so large as to indicate an error in calculation; (v) that a difficulty involved in the pursuit of method (a) to smaller working interval for the open-range section is that $f^{(4)}(\xi)$ will at the same time increase rapidly; (vi) the methods of (b) and (c) are sometimes known as 'subtracting out the singularity' and 'dividing out the singularity', respectively.

Problem 5. Evaluate π^2 to 6D from the series $\pi^2/6 = \sum_{m=1}^{\infty} (1/m^2)$.

The process of summing terms of a series is related to that of integration. In general we need distinct formulae for the two types of process, but when the corrected form of the trapezoidal rule is expressed in terms of derivatives of the integrand (in place of finite differences) we obtain a formula serving both processes, and this is the *Euler-Maclaurin formula**,

$$\int_{x_0}^{x_0+nh} f(x)dx = h\left(\frac{1}{2}f_0 + f_1 + f_2 + \ldots + f_{n+1} + \frac{1}{2}f_n\right) - \frac{1}{12}h^2(f'_n - f'_0) +$$
$$+ \frac{1}{720}h^4(f'''_n - f'''_0) - \frac{1}{30240}h^6(f^{(5)}_n - f^{(5)}_0) +$$
$$+ \frac{1}{1209600}h^8(f^{(7)}_n - f^{(7)}_0) + \ldots$$

For the summation of terms of a series the Euler-Maclaurin formula can be rewritten:
$$f_0 + f_1 + \ldots + f_n$$
$$= \frac{1}{h}\int_{x^0}^{x_0+nh} f(x)\,dx + \frac{1}{2}(f_n + f_0) + \frac{1}{12}h(f'_n - f'_0) - \frac{1}{720}h^3(f'''_n - f'''_0) +$$
$$+ \frac{1}{30240}h^5(f^{(5)}_n - f^{(5)}_0) - \frac{1}{1209600}h^7(f^{(7)}_n - f^{(7)}_0) \ldots) +$$

---
* For a derivation of the Euler-Maclaurin formula see for example, E. T. Whittaker and G. N. Watson, *A course of modern analysis* (Cambridge University Press), chapter 7, §3; it can also be derived from the trapezoidal formula for an extended range by substituting for finite differences in terms of derivatives.

DIFFERENTIATION AND INTEGRATION

Hence, when the general term of a series can be written $f(m)$, where $f(x)$ is a function possessing the necessary derivatives and which can be integrated, we can form sums of terms of the series. In the present case $f(m) = m^{-2}$; thus $f(x) = x^{-2}$ and $f^{(k)}(x) = (-1)^k (k+1)!/x^{k+2}$. To sum the series take $h=1$ and $x=1$, when

$$\sum_1^n \frac{1}{m^2} = \int_1^{n+1} \frac{dx}{x^2} + \frac{1}{2}\left(\frac{1}{(n+1)^2} + \frac{1}{1^2}\right) + \frac{1}{12}\left(-\frac{2}{(n+1)^3} + \frac{2}{1^3}\right)$$
$$- \frac{1}{720}\left(-\frac{24}{(n+1)^5} + \frac{24}{1^5}\right) + \ldots ;$$

and if we take the limit of the sum as $n \to \infty$,

$$\sum_1^\infty \frac{1}{m^2} = \frac{1}{1} + \frac{1}{2.1^2} + \frac{1}{6.1^3} + \frac{1}{30.1^5} - \ldots .$$

These terms decrease very slowly, and we should have to accumulate far too many for the required accuracy. This is because $f^{(k)}(1) = (-1)^k (k+1)!$, and it is apparent that a much more effective process can be obtained if we sum the first few terms directly and apply the Euler-Maclaurin formula to the remaining terms. Thus

$$\sum_{m=10}^\infty \frac{1}{m^2} = \frac{1}{10} + \frac{1}{2.10^2} + \frac{1}{6.10^3} - \frac{1}{30.10^5} + \frac{1}{42.10^7} - \ldots$$
$$= \cdot 1000000_0 + \cdot 0050000_0 + \cdot 0001666_7 - \cdot 0000003_3 + \cdot 0000000_0$$
$$= \cdot 1051663_4.$$

By direct addition

$$\sum_1^9 \frac{1}{m^2} = 1 \cdot 5397677_3$$

and hence $\pi^2 = 6 \times 1 \cdot 6449340_7 = 9 \cdot 869604$. The obvious check is by repeating the process, taking a different number of initial terms for direct addition, and this must be repeated until two successive answers agree.

Note: (i) that this form of the Euler-Maclaurin formula does not imply that it is an infinite series, and for many integrands it will prove to be an asymptotic approximation in which the accuracy improves with each successive term only up to a certain stage and thereafter decreases progressively; (ii) that the Euler-Maclaurin formula can be used in its first form for integration of a specified function provided that derivatives can be formed and evaluated with reasonable ease; (iii) that we must retain *eight* decimal digits if after adding $9+4$ terms (each with possible round-off of ·5 in the last digit) and multiplying by 6 we are to achieve a result correct to 6D; (iv) that to obtain 6D accuracy in π^2 by direct addition of terms we should have to add some twelve million terms of the series!

4.4. Double and repeated integration

Double (and other multiple) integrals can be evaluated by use of the foregoing general techniques applied to a repeated integral. Thus

$$\int_{-h}^{h}\int_{-h}^{h} f(x, y)dxdy = \int_{-h}^{h}\left\{\int_{-h}^{h} f(x, y)dx\right\}dy,$$

and if we take $F(y) = \int_{-h}^{h} f(x, y)dx$, we have from Simpson's rule

$$\int_{-h}^{h}\int_{-h}^{h} f(x, y)dxdy$$

$$\simeq \frac{1}{3}h\{F(-h) + 4F(O) + F(h)\}$$

$$= \frac{1}{3}h\left\{\int_{-h}^{h} f(x, -h)dx + 4\int_{-h}^{h} f(x, 0)dx + \int_{-h}^{h} f(x, h)dx\right\}$$

$$\simeq \frac{1}{3}h\left\{\frac{1}{3}h[f_{-1-1} + 4f_{0-1} + f_{1-1}] + \frac{4}{3}h[f_{-10} + 4f_{00} + f_{10}] + \frac{1}{3}h[f_{-11} + 4f_{01} + f_{11}]\right\}$$

$$= \frac{1}{9}h^2\{f_{11} + f_{-11} + f_{-1-1} + f_{1-1} + 4f_{10} + 4f_{01} + 4f_{-10} + 4f_{0-1} + 16f_{00}\},$$

DIFFERENTIATION AND INTEGRATION

where $f_{ij} \equiv f(ih, jh)$. In principle this is a perfectly satisfactory form of Simpson's rule for a double integral, in which the value of the integral is given in terms of tabular values at the junctions of a square mesh of side-length $2h$, as indicated in figure 4. However, it will obviously be laborious to apply, and we should prefer a formula with fewer terms on the right-hand side if this could be obtained without undue loss of accuracy.

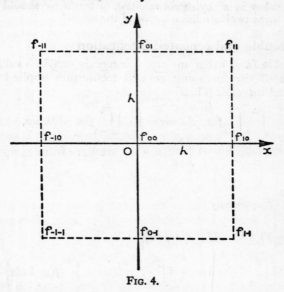

Fig. 4.

One simpler formula, in which the x, y symmetry is preserved but from which the relatively distant 'corner terms' $f_{11}, f_{-11}, f_{-1-1}, f_{1-1}$ are omitted, is of the general form

$$\int_{-h}^{h}\int_{-h}^{h} f(x, y)dxdy \simeq af_{00} + b(f_{10} + f_{01} + f_{-10} + f_{0-1}),$$

where a and b are constants to be determined. These con-

4.4 DOUBLE AND REPEATED INTEGRATION

stants can be found easily if we make use of the idea of polynomial approximation in a slightly different way, and in particular of quadratic approximation which we have already found to be effective in integration. If the formula that we are constructing is to correspond with quadratic matching of the integrand $f(x, y)$ then it will integrate correctly (i.e. without *any* error) all quadratic functions of x and y, including of course linear functions and constants. Moreover, since integration is a linear process, we can be sure that all quadratic functions will be integrated correctly if this is true independently for x^2, x (and hence by symmetry for y^2, y), xy and 1.

When $f(x, y) = 1$ we have from the formula

$$\int_{-h}^{h}\int_{-h}^{h} 1 \, dx \, dy \eqsim a + 4b;$$

this double integral can be evaluated directly and has the value $4h^2$, and hence the formula integrates constants correctly if $a + 4b = 4h^2$. For $f(x, y) = x$,

$$0 = \int_{-h}^{h}\int_{-h}^{h} x \, dx \, dy \eqsim a \times 0 + b \times 0,$$

so that the formula integrates x, and indeed xy and all odd powers of x or y (and products odd in x or y) correctly. For $f(x, y) = x^2$,

$$(2h^3/3)2h = \int_{-h}^{h}\int_{-h}^{h} x^2 \, dx \, dy \eqsim a \times 0 + b \times 2h^2,$$

and the formula will integrate x^2 correctly (and also y^2, by symmetry) if $b = 2h^2/3$. Hence the formula will integrate all quadratic functions correctly if $a = 4h^2 - 4b = 4h^2/3$, and we have the double integration formula based on quadratic matching,

$$\int_{-h}^{h}\int_{-h}^{h} f(x, y) \, dx \, dy \eqsim \frac{2}{3} h^2 (2f_{00} + f_{10} + f_{01} + f_{-10} + f_{0-1})$$

$$= \tfrac{1}{6}(\text{area of element})(2f_{00} + f_{10} + f_{01} + f_{-10} + f_{0-1}).$$

DIFFERENTIATION AND INTEGRATION

Other formulae of this general type can be obtained in a similar way, and these can be tailored to fit any special requirements of the problem in hand; indeed, the derivation above has been given in detail to demonstrate the construction of a formula as well as for its own importance. In each case we can calculate a correction term provided that a Taylor series (in x and y) is available, and the correction terms for the formulae of this section are

$$-(h^6/45)\{f_{xxxx}(\xi,\eta)+f_{yyyy}(\xi,\eta)\}$$

and

$$-(h^6/45)\{f_{xxxx}(\xi,\eta)-5f_{xxyy}(\xi,\eta)+f_{yyyy}(\xi,\eta)\},$$

respectively, where (ξ,η) is a point of the square domain of integration and $f_{xxxx} \equiv \partial^4 f/\partial x^4$, etc.

4.5. Gaussian quadrature

Our previous integration formulae have all been based on the use of equally spaced tabular values of the integrand. Such formulae are convenient for integration from constant interval tables, and also for integrands of given functional form (as the tabular points can be chosen for easy calculation), but they are not the most accurate type of formula obtainable. If instead we specify in advance the *number of tabular values* that we wish to incorporate in an integration formula *but not their position*, we can construct a formula by finding the tabular positions within the range of integration for the most accurate integration formula obtainable with the specified number of values of the integrand. This approach is due to Gauss.

Thus, the most general form of two-point formula of this type is

$$\int_{-h}^{h} f(x)dx \simeq h\{af(\alpha h)+bf(\beta h)\},$$

where the *weighting constants* a and b, and the *position factors* α and β are constants to be determined. The approximation symbol $\simeq$ has been used in anticipation of the fact

4.5 GAUSSIAN QUADRATURE

that we shall now obtain formulae based on polynomial approximation (using the device of the previous section). The simplest formula of this type* is symmetrical about $x = 0$ and uses values of the integrand calculated at points $\pm \alpha h$; for this case $\beta = -\alpha$. We now obtain the values of a, b, and α by stipulating that the formula will integrate 1, x, and x^2 (and hence all quadratic functions) without error.

The constant 1 will be integrated correctly if $2h = h(a+b)$; and x correctly if $0 = h(a\alpha h - b\alpha h)$; and x^2 correctly if $2h^3/3 = h(a\alpha^2 h^2 + b\alpha^2 h^2)$. Hence $a = b = 1$, and $\alpha^2 = 1/3$; and we have the *Gauss two-point integration formula*

$$\int_{-h}^{h} f(x)dx \simeq h\{f(h/\sqrt{3}) + f(-h/\sqrt{3})\};$$

this formula (like Simpson's rule) integrates powers up to x^3 (and all odd powers of x) correctly.

The Gauss two-point formula integrates quadratics correctly and corresponds to quadratic representation of the integrand $f(x)$. The error may be calculated using Taylor expansions, and the *leading correction term* is found to be $+(h^5/135)f^{(4)}(\xi)$, which is very similar in magnitude to that for Simpson's rule but is of opposite sign. It should be noted that the Gauss two-point formula achieves as good accuracy as Simpson's rule in spite of the fact that it employs only two instead of three tabular values; this is the advantage of Gauss type formulae, but it must be compared with the disadvantage that these values must be determined at awkward points. Thus when we are working by hand the inconvenience of evaluating the integrand at the Gauss points will more than offset the improved accuracy in general and Simpson's rule (say) will be more effective; but when we are working with a high-speed computing machine which can handle any tabular points

---

* For special applications we could easily develop non-symmetrical formulae of this type.

DIFFERENTIATION AND INTEGRATION

with equal ease the Gauss formulae will have real advantages.

The *Gauss three-point integration formula*

$$\int_{-h}^{h} f(x)dx$$
$$= \frac{1}{9}h\left\{5f\left(-h\sqrt{\frac{3}{5}}\right) + 8f(0) + 5f\left(h\sqrt{\frac{3}{5}}\right)\right\} + (h^7/15750)f^{(6)}(\xi)$$

can be obtained in a similar way, as the formula of type $h\{af(-\alpha h) + af(\alpha h) + bf(0)\}$ corresponding to cubic approximation to $f(x)$.

EXERCISES ON CHAPTER FOUR

1. Derive the half-way differentiation formula

$$hf'_{1/2} \simeq \delta f_{1/2} - \frac{1}{24}\delta^3 f_{1/2} + \frac{3}{640}\delta^5 f_{1/2} - \frac{5}{7168}\delta^7 f_{1/2} + O(h^9).$$

Find $f'(\cdot 55)$ from the following table for $f(x) = e^{-x}$, and check the result by interpolation.

x	·3	·4	·5	·6	·7	·8
$f(x)$	·74082	·67032	·60653	·54881	·49659	·44933

2. Obtain the differentiation formulae, based on tabular values,

$$2hf_0' = f_1 - f_{-1} - \tfrac{1}{3}h^{2\prime\prime\prime}f(\xi),$$
$$12hf_0' = -f_2 + 8f_1 - 8f_{-1} + f_{-2} + \frac{1}{30}h^4 f^{(5)}(\xi).$$

3. (i) Derive the *three-eighths rule* for integration

$$\int_0^{3h} f(x)dx \simeq \tfrac{3}{8}h(f_0 + 3f_1 + 3f_2 + f_3)$$

using the matching polynomial $P_3(x)$, and show that the correction term can be written as $-\frac{3}{80}h^5 f^{(4)}(\xi)$.

(ii) Derive *Weddle's rule*

$$\int_0^h f(x)dx \simeq \frac{3}{10}h(f_0 + 5f_1 + f_2 + 6f_3 + f_4 + 5f_5 + f_6)$$

EXERCISES

from the polynomial $P_6(x)$. (To obtain this form a small term must be neglected.) Find the correction term.

(iii) Compare the accuracy of *Dufton's rule*

$$\int_0^{10h} f(x)dx \simeq \frac{5}{2}h(f_1+f_4+f_6+f_9)$$

with that of the trapezoidal rule and Simpson's rule.

4. Derive the integration formula for use with divided differences

$$\int_{x_0}^{x_1} f(x)dx \simeq (x_1-x_0)f(x_0)+\tfrac{1}{2}(x_1-x_0)^2[x_0x_1]-\tfrac{1}{6}(x_1-x_0)^3[x_0x_1x_2],$$ and

show that the correction term is $\int_{x_0}^{x_1}(x-x_0)(x-x_1)(x-x_2)[xx_0x_1x_2]dx$.

5. Evaluate $\int_0^{1.5} f(x)dx$ from the following table for the error function
$f(x)=\sqrt{2/\pi}e^{-x^2/2}$, comparing the results obtained using the trapezoidal rule, Simpson's rule, the three-eighths rule and Weddle's rule.

x	$f(x)$	x	$f(x)$	x	$f(x)$
·000	·79788	·625	·65632	1·250	·36530
·125	·79168	·750	·60227	1·375	·31002
·250	·77334	·875	·54411	1·500	·25903
·375	·74371	1·000	·48394		
·500	·70413	1·125	·42375		

6. Integrate $1/(10+x^2+y^2)$ correct to 5D over the square region with corners $(\pm 1, \pm 1)$. [Carry out the integration with the region divided into one, four, and then nine elementary squares, and compare the values obtained from these three different working subdivisions of the region.]

7. Derive the repeated integration formula

$$\int_0^h dx \int_0^x f(t)dt \simeq \tfrac{1}{2}h^2\left(f_0+\tfrac{1}{3}\mu\delta f_0+\tfrac{1}{12}\delta^2 f_0-\tfrac{7}{180}\mu\delta^3 f_0-\tfrac{1}{240}\delta^4 f_0+\ldots\right).$$

Hence deduce the repeated integration formula

$$\int_{-h}^{h} dx \int_0^x f(t)dt \simeq h^2\left(f_0+\tfrac{1}{12}\delta^2 f_0-\tfrac{1}{240}\delta^4 f_0+\tfrac{31}{60480}\delta^6 f_0+\ldots\right).$$

8. Construct a double integration formula of type

$$\int_{-h}^{h} dx \int_{-h}^{h} f(x,y)\,dy \simeq a f_{00}+b(f_{11}+f_{-11}+f_{-1-1}+f_{1-1}).$$

What kind of polynomial matching does it correspond to? Give a Taylor series estimate of the correction term.

CHAPTER FIVE

Method of Least Squares

Finite difference processes may be used effectively with tables of well-represented functions because the corresponding differences decrease rapidly with increasing order. Thus higher differences of mathematical tables (with reasonably short interval) tend to be small and regular in behaviour; and indeed we detect occasional blunders by the relatively large disturbances they cause in the otherwise systematic behaviour of the difference table. The same processes will generally be ill-suited for use with tables of experimental results, as these will include random fluctuations that often dominate the higher differences. *Most* experimental values will be in error (by an amount which exceeds normal round-off), and the effects of these errors in the higher differences will be so intertwined that there is no hope of detecting (let alone of correcting) them by the method of §2.5. We must develop some other way of handling experimental results.

So far we have selected our approximating polynomials to agree exactly with the specified tabular values over a chosen range of tabular points; thus we tolerate some error of approximation between tabular points, though none at the points themselves. But when the tabular values include random errors there is no advantage to be gained from fitting them precisely, and we would prefer approximating polynomials that follow the general trend of a table without reproducing all the local fluctuations. This relaxation of the accuracy with which we fit individual tabular values involves a reduction in constraint on the approximating polynomial and a consequent increase of range over which a polynomial of given degree may be used in approximation. Matching processes which involve some averaging of small-scale local fluctuations are known as *curve-fitting*.

5.1. Fitting straight lines to data

It is often obvious from graphical inspection that the measured values, say $y_i = y(x_i)$, obtained from an experiment lie roughly along a line. These results apparently define a linear relationship of the form

$$y \simeq a_0 + a_1 x = \overline{P}_1(x),$$

where the linear representation $\overline{P}_1(x)$ is not a Newton polynomial (which would have zero error of approximation at tabular points) but is some sort of averaged representation; we want a procedure for selecting such a 'line of best fit' or 'average line' through the points. It should be noted that

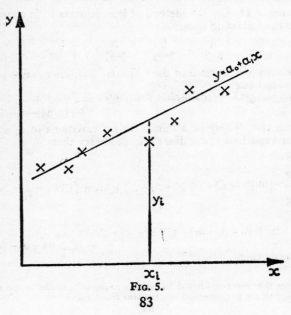

Fig. 5.

METHOD OF LEAST SQUARES

there is no ultimate best line, unless the points are collinear, and we shall obtain *different* 'best' lines according to the criteria of selection we adopt. A good line will pass close to most points, and we can select such a line if we adjust a_0 and a_1 until *the sum of the squares of the residual distances**
$y_i - (a_0 + a_1 x_i)$ *is a minimum*. The use of (distance)^2 rather than simple distance provides a weighting towards distant points so that points already near the line exert relatively less influence and contributions from points on opposite sides of the line reinforce. This *criterion of least squares* will ensure that the distribution of points as a whole lies closely about the best line; it is not the only criterion that could be used, but it is convenient in many applications. The *method of least squares* can best be illustrated by an example of linear approximation.

Example 11. Use the method of least squares to fit a straight line to the following data.

x	0	1	2	3	4	5
y	0·82	3·46	5·92	8·57	11·07	13·64

The sum of squares S of the residual distances relative to the line $y = a_0 + a_1 x$ is
$$S = (0·82 - a_0)^2 + (3·46 - a_0 - a_1)^2 + (5·92 - a_0 - 2a_1)^2 + \ldots$$
$$+ (13·64 - a_0 - 5a_1)^2.$$

In order that S may be a minimum for variations of a_0 and a_1 (i.e. for variations of the line) it is necessary that

$$0 = \frac{\partial S}{\partial a_0}$$
$$= -2\{(0·82 - a_0) + (3·46 - a_0 - a_1) + \ldots + (13·64 - a_0 - 5a_1)\},$$
$$0 = \frac{\partial S}{\partial a_1}$$
$$= -2\{(3·46 - a_0 - a_1) + 2(5·92 - a_0 - 2a_1) + \ldots$$
$$+ 5(13·64 - a_0 - 5a_1)\};$$
and hence that

---
* Note that these residual distances are measured parallel to the *y*-axis or ease, but are proportional to distances from a line

5.2 APPROXIMATION BY LEAST SQUARES

$$6a_0 + 15a_1 = 43 \cdot 48$$
$$15a_0 + 55a_1 = 153 \cdot 49.$$

Thus the line of best fit in the least squares sense is

$$y \simeq 0 \cdot 85 + 2 \cdot 56 x.$$

Note: (i) that we have not proved that this line corresponds to a *minimum* of S, though this is not difficult to do; (ii) that the approximate values at tabular points are $0\cdot 85$, $3\cdot 41$, $5\cdot 97$, $8\cdot 53$, $11\cdot 09$, $13\cdot 65$; (iii) that a measure of the effectiveness of the approximation is provided by the root mean square of the remaining residuals

$$\{\tfrac{1}{6}[(\cdot 03)^2 + (\cdot 05)^2 + (\cdot 05)^2 + (\cdot 04)^2 + (\cdot 02)^2 + (\cdot 01)^2]\}^{1/2} = 0\cdot 037;$$

(iv) that we cannot strictly determine the accuracy of the approximation as we do not know the correct tabular values; (v) that we can fit a Newton polynomial $P_1(x)$ to just two tabular points, but that a least squares polynomial $\overline{P}_1(x)$ can be fitted to any number of points.

5.2. Polynomial approximation by least squares

In very much the same way we can find polynomial approximations $\overline{P}_n(x)$ of any degree for given data; the choice of degree is in no way regulated by the number of data points, though the accuracy of approximation will depend on our choice. The procedure can conveniently be described in terms of the algebraically simple case of fitting a parabola $\overline{P}_2(x) = a_0 + a_1 x + a_2 x^2$ to the $n+1$ tabular values $f(x_i)$ ($n > 2$, and $i = 0, 1, \ldots n$).

We must choose the coefficient parameters a_0, a_1, and a_2 so as to obtain a minimum value for the sum of residuals

$$S = \sum_{i=0}^{n} \{f(x_i) - a_0 - a_1 x_i - a_2 x_i^2\}^2.$$

A set of necessary conditions for S to be a minimum is that $\partial S / \partial a_k = 0$ for $k = 0, 1, 2$; and hence

$$-2 \sum_{1}^{n} \{f(x_i) - a_0 - a_1 x_i - a_2 x_i^2\} x_i^k = 0 \qquad k = 0, 1, 2.$$

METHOD OF LEAST SQUARES

If we take $s_k = \sum_{i=0}^{n} x_i^k$ and $v_k = \sum_{i=0}^{n} x_i^k f_i$, these equations can be written

$$s_0 a_0 + s_1 a_1 + s_2 a_2 = v_0$$
$$s_1 a_0 + s_2 a_1 + s_3 a_2 = v_1$$
$$s_2 a_0 + s_3 a_1 + s_4 a_2 = v_2,$$

and are known as the *normal equations* of the system. It can be shown* that the normal equations always have a unique solution for a_0, a_1, a_2 and that this solution does correspond to a minimum of S.

The derivation of the normal equations should be laid out systematically, as in the following table:

1	x	x^2	x^3	x^4	f	xf	$x^2 f$
1	x_0	x_0^2	x_0^3	x_0^4	f_0	$x_0 f_0$	$x_0^2 f_0$
1	x_1	x_1^2	x_1^3	x_1^4	f_1	$x_1 f_1$	$x_1^2 f_1$
1	x_2	x_2^2	x_2^3	x_2^4	f_2	$x_2 f_2$	$x_2^2 f_2$
.	.	.	.	.	.	.	.
1	x_n	x_n^2	x_n^3	x_n^4	f_n	$x_n f_n$	$x_n^2 f_n$
s_0	s_1	s_2	s_3	s_4	v_0	v_1	v_2

The values x_i and f_i are given, and may be entered forthwith; their powers and products are then calculated and entered, and the columns are added to obtain s and v sums.

Example 12. Fit a least squares polynomial to the following experimental data (which may be in error to the extent of about ·02 per value):

x	0	1	2	3	4	5	6
y	−·01	·37	·71	·98	1·28	1·52	1·69

A substantial saving in labour can be obtained from the change of origin $\bar{x} = x - 3$: in terms of $\bar{x}$ the tabular points are

---

* See Milne, *Numerical Calculus* (Princeton, 1949).

5.2 APPROXIMATION BY LEAST SQUARES

symmetrically arranged about $\bar{x}=0$, and if the tabular interval is uniform the odd sums are zero, i.e. $s_1=0$, $s_3=0$, ...

We obtain the normal equations for a *linear* approximation as follows:

1	$\bar{x}$	$\bar{x}^2$	y	$\bar{x}y$
1	-3	9	$-\cdot 01$	$\cdot 03$
1	-2	4	$\cdot 37$	$-\cdot 74$
1	-1	1	$\cdot 71$	$-\cdot 71$
1	0	0	$\cdot 98$	0
1	1	1	$1\cdot 28$	$1\cdot 28$
1	2	4	$1\cdot 52$	$3\cdot 04$
1	3	9	$1\cdot 69$	$5\cdot 07$
7	0	28	$6\cdot 54$	$7\cdot 97$

Hence
$$7a_0 + 0a_1 = 6\cdot 54$$
$$0a_0 + 28a_1 = 7\cdot 97,$$

and
$$y \simeq 0\cdot 93_4 + 0\cdot 28_5 \bar{x}.$$

A measure of the effectiveness of this approximation is provided by the r.m.s. (root mean square) of the remaining residuals; the remaining rms residual is $0\cdot 06$, and as this is substantially larger than the anticipated experimental error it appears that this linear approximation is inadequate.

The normal equations for the quadratic approximation follow as:

1	$\bar{x}$	$\bar{x}^2$	$\bar{x}^3$	$\bar{x}^4$	y	$\bar{x}y$	$\bar{x}^2 y$
1	-3	9	-27	81	$-\cdot 01$	$\cdot 03$	$-\cdot 09$
1	-2	4	-8	16	$\cdot 37$	$-\cdot 74$	$1\cdot 48$
1	-1	1	-1	1	$\cdot 71$	$-\cdot 71$	$\cdot 71$
1	0	0	0	0	$\cdot 98$	0	0
1	1	1	1	1	$1\cdot 28$	$1\cdot 28$	$1\cdot 28$
1	2	4	8	16	$1\cdot 52$	$3\cdot 04$	$6\cdot 08$
1	3	9	27	81	$1\cdot 69$	$5\cdot 07$	$15\cdot 21$
7	0	28	0	196	$6\cdot 54$	$7\cdot 97$	$24\cdot 67$

$$7a_0 + 0a_1 + 28a_2 = 6\cdot 54$$
$$0a_0 + 28a_1 + 0a_2 = 7\cdot 97$$
$$28a_0 + 0a_1 + 196a_2 = 24\cdot 67.$$

METHOD OF LEAST SQUARES

Hence
$$y \simeq 1\cdot00_5 + 0\cdot28_5\,\bar{x} - 0\cdot01_8\,\bar{x}^2$$
$$= -0\cdot01_2 + 0\cdot39_3\,x - 0\cdot01_8\,x^2.$$

The rms residual error is now $0\cdot01$, which is well within the range of possible experimental error and must be regarded as satisfactory.

Note: (i) that we now rely on our judgment in choosing the degree of approximating polynomial that is appropriate, and there is nothing to prevent us from finding a 'best straight line' through points obviously ill-suited for linear approximation; (ii) that the root mean square residual is a valuable way of assessing an approximation by least squares; (iii) that there is an *effective* reduction by one in the number of normal equations to be solved when the tabular points are arranged symmetrically; (iv) that the third decimal digits are needed as subscripts (but cannot genuinely be written as 'correct digits') in order to preserve accuracy.

Experimental results are often plotted on log-log or log-plain graph paper to bring out relationships of power law and exponential type. The same procedure can be carried out much more accurately using least-square linear approximation. In the case $y = ax^n$, we have $\log y = \log a + n \log x$, and if we put $X = \log x$, $Y = \log y$, and $A = \log a$, we have,

$$Y = A + nX;$$

A and n can be determined now by a least-squares process applied to the values $\log y_i$ at points $\log x_i$. Similarly, if $y = ae^{bx}$, we take $Y = \log y$ and $A = \log a$ to give $Y = A + bx$, and a least-squares process can be applied to the values $\log y_i$ at points x_i.

It should be emphasized that the least-squares approach is bound to yield an approximation,* though we shall get a useless approximation if we select the degree of the approximating polynomial foolishly. It follows that we must always test the significance of an approximation by a final calculation of the rms residual, or in some other way.

---

* Compare the comments on semi-empirical theories of §3.7 (iii).

5.2 APPROXIMATION BY LEAST SQUARES

In some applications the tabular values may be regarded as of unequal importance or *weight*; in such cases a *weight function* $w(x_i)$ can be associated with each tabular point, and the least-squares process will then be carried out on the weighted sum

$$S = \sum w(x_i)\{f(x_i) - \bar{P}_n(x_i)\}^2.$$

Problem 6. Find a least squares quadratic approximation for the continuous function e^x over the range $(-1, 1)$.

We might select an arbitrary set of tabular points spanning the range $(-1, 1)$ and apply the methods of this section to the corresponding set of values for e^x. But it will be obviously superior to use the known values of e^x right across the range by using integration in place of the summation process necessary for sets of discrete values. Thus, in place of the sum of squares of the residuals we take the integral

$$S = \int_{-1}^{1} \{e^x - (a_0 + a_1 x + a_2 x^2)\}^2 dx,$$

which is a function of a_0, a_1 and a_2. A necessary set of conditions for a minimum of the function $S(a_0, a_1, a_2)$ is that $\partial S/\partial a_k = 0$ for $k = 0, 1, 2$: hence

$$0 = \frac{\partial S}{\partial a_0} = -2 \int_{-1}^{1} \{e^x - (a_0 + a_1 x + a_2 x^2)\} dx,$$

$$0 = \frac{\partial S}{\partial a_1} = -2 \int_{-1}^{1} \{e^x - (a_0 + a_1 x + a_2 x^2)\} x \, dx,$$

$$0 = \frac{\partial S}{\partial a_2} = -2 \int_{-1}^{1} \{e^x - (a_0 + a_1 x + a_2 x^2)\} x^2 \, dx.$$

After integration, we obtain the set of normal equations

$$3a_0 + a_2 = 3 \cdot 53$$
$$a_1 = 1 \cdot 10$$
$$5a_0 + 3a_2 = 6 \cdot 59,$$

and the quadratic approximation

$$y \simeq 1 \cdot 00 + 1 \cdot 10 x + 0 \cdot 53 x^2.$$

Note: (i) that the numerical value for e has been substituted immediately as we are concerned with *numerical* solutions;

(ii) that the error does not exceed about ·03 units except at the very ends of the range, and that the rms residual error is ·03; (iii) that the approximation differs from the first three terms of the Taylor expansion near $x=0$, but is more accurate over the specified range *as a whole*; (iv) that this process of matching a continuously defined function is known as *continuous approximation by least squares*.

5.3. Data smoothing

Raw experimental results provide our only source of contact with the physical world; we should on no account 'adjust' these results capriciously, either to improve an 'apparent pattern of dependence' (which may not be a genuine one, in any case) or to provide a better level of agreement with a favoured theory. Experiments which yield consistently poor results may need refining, or they may be based on an inadequate conception of the phenomena involved; in either case we can improve the apparent quality of the results by *smoothing processes*, but we will not improve their significance (though we may appear to do so!). Thus data smoothing is *not* an alternative to good experimental technique. However, it must be appreciated that even good experiments yield tables of results with enough experimental scatter to produce sizable spurious fluctuations in the higher differences. Some initial smoothing may be necessary before such tables are used in numerical calculations, especially for processes like differentiation which are very sensitive to local error.

In smoothing a table of values we do not attempt to find a single approximating polynomial for the whole table; instead we use repeated local matching with low-degree polynomials, and each polynomial is used only to adjust a single central value (except at the ends of the range). Thus we assume that the table is *locally* suitable for approximation by polynomials of low degree, apart from the experimental scatter, but we make no assumption about large-scale approximation of the table.

5.3 DATA SMOOTHING

A wide range of formulae is available for smoothing data at uniform interval; we shall derive a quadratic five-point formula. We shall suppose that a run of five values is sufficient to correct for random errors in the measured value of the *central* member, and we shall take the tabular interval as the unit of length and the central point as origin. The normal equations for fitting a parabola $a_0 + a_1 x + a_2 x^2$ to the values $f_{-2}, f_{-1}, f_0, f_1, f_2$ by least-squares are then

$$5a_0 + 10a_2 = \sum_{-2}^{2} f_i$$

$$10a_1 = \sum_{-2}^{2} x_i f_i$$

$$10a_0 + 34a_2 = \sum_{-2}^{2} x_i^2 f_i.$$

For use in the interior of the table we want only a corrected value for the central point, that is the value a_0; and we have on solution

$$35a_0 = 35f_0 - 3(f_{-2} - 4f_{-1} + 6f_0 - 4f_1 + f_2)$$
$$= 35f_0 - 3\delta^4 f_0.$$

Hence the corrected value* at $x = 0$ is $y_0 \simeq f_0 - 3\delta^4 f_0/35$. For use near the head or foot of a table

$$y_{-2} \simeq f_{-2} - \frac{1}{70}\delta^4 f_0, \quad y_2 \simeq f_2 - \frac{1}{70}\delta^4 f_0,$$

$$y_{-1} \simeq f_{-1} + \frac{2}{35}\delta^4 f_0, \quad y_1 \simeq f_1 + \frac{2}{35}\delta^4 f_0.$$

In selecting the degree n for the smoothing process we should take n small so that individual calculations are easy, and yet we must so judge our choice that differences of the

* For many purposes it will be sufficient to use the approximate form $y_0 \simeq f_0 - \delta^4 f_0/12$.

METHOD OF LEAST SQUARES

true function of order greater than n are small. In fact the choice may often be made from the fact that the first n differences are reasonably regular, but the $(n+1)$th differences are irregular and have mean value near zero. The amount of smoothing increases with the number of values fitted, and decreases with the degree of the approximating polynomial. Obviously there is great scope for *judgment* on the part of the computer as to what is random and what genuine; sometimes the application of successive smoothing processes may be justified, but repeated smoothing will ultimately reduce *any* set of data to a straight line.

EXERCISES ON CHAPTER FIVE

1. Use the method of least squares to find the best straight line fitting the data

x	0	1	2	3	4	5	6	7	8	9	10
y	15	14	13	13	12	10	6	8	5	4	2;

2. Find the best curve of type $y = ax^n$ fitting the data

x	0·13	0·33	1·08	3·65	4·98
y	3	4	6	9	10.

Note on Operational Methods

Most readers will have some familiarity with operational methods through their use in solving linear differential equations with constant coefficients. It is possible in a purely formal way to dissociate the operation of differentiation (or integration) from any particular function $f(x)$ and to establish a *formal algebra* for *the operator* $D = d/dx$, which is similar to (though not identical with) the algebra of real numbers.

In precisely the same way we can regard the operations of differencing as distinct from their application to a particular function, in which case Δ, ∇ and δ may be regarded as *differencing operators* and μ as an *averaging operator*. A formal algebra can be established in terms of these and other operators, and this operational algebra can be used in deriving a very wide range of formulae for numerical processes. Operational methods are used in many books because they allow of such easy development of formulae, but they have not been used here as they tend to leave the significance of the various steps obscure to the inexperienced reader. We have advanced more slowly, in the hope that the reader will carry a clear understanding of the reasons behind our development of processes and of our attitude towards numerical work; he can then take up the operational approach with help from a more advanced text.

Answers and Notes on Exercises

Chapter One

1. (i) 3, 6; about 1 in 2×10^6. (ii) 30, 3; about 0·05 per cent. (iii) 0, **5**; about 1 in 60,000. (iv) 4, 3; about 0·4 per cent, and note that the final zero is needed to define this accuracy.

2. 1·38 and 1·37; note that the rounding error can (just) exceed half a unit in the last digit when a number is rounded digit by digit.

3. (i) 0·5; actually $0·47 + 7 \times ·005$, and hence the maximum error in the second decimal from round-off is $\pm 3\frac{1}{2}$, so that this digit is uncertain (the probable error is considerably smaller, but exceeds ·005). (ii) 83·5; maximum error is about $(9·37 + 8·91) \times ·005 = 0·09$, and the second decimal in 83·49 is unreliable and the error *could* affect the first decimal. (iii) 25·6; *maximum* rounding error is 0·048 after division and the second and third decimals are unreliable (clearly the answer cannot be as accurate as the denominator 0·279!). (iv) 17·760; the decimal accuracy of these numbers varies and as the middle number is known only within ·0005 there is no justification for retaining more than three decimals, maximum error is about ·0006. *Note* that maximum errors can be found using the binomial expansion, as

$$(a \pm \epsilon)/(b \pm \delta) = \frac{a}{b}\left(1 \pm \frac{\epsilon}{a}\right)\left(1 \pm \frac{\delta}{b}\right)^{-1} = \frac{a}{b}\left\{1 \pm \left(\frac{\epsilon}{a} + \frac{\delta}{b}\right) + \ldots\right\}.$$

4. (i) 0·693144 with maximum rounding error of 1 in last digit, (ii) 0·693143 with maximum rounding error of 4 in last digit; this illustrates the value of carrying an extra working digit for some calculations.

5. (i) $1 + x$, for the range $-·1 < x < ·1$; (ii) $1 + x + \frac{1}{2}x$, for the range $-·3 < x < ·3$; (iii) $1 + x + \frac{1}{2}x^2 + \frac{1}{6}x^3$, for the range $-·5 < x < ·5$.

Chapter Two

2. Compare with formula for differentiation of products, etc.

3. Use induction.

4. Third differences are constant: hence data fitted by a cubic which is obtained from an interpolation formula as $75 + 521x/6 - 25x^2/2 + 2x^3/3$.

5. $f(1·5) = 1·142892$. Transposition is a common blunder; the adjusted fourth differences are about as regular as can be expected, allowing for round-off.

ANSWERS AND NOTES ON EXERCISES

Chapter Three

1. Use Newton-Gregory interpolation with forward differences near head of table and backward differences near foot. Differencing of final table will uncover any errors, so that there is no need to give solution here.

2. (i) Suitable for linear interpolation as $\delta^2 f \leqslant 1$ over whole table. (ii) Not suitable, as $\delta^2 f$ rises from 15 to 29. Subtabulation to half-interval will reduce second differences by factor of about $1/2^2$; they are still too large to be neglected. Subtabulation to one-third interval will reduce second differences by factor of about $1/3^2$, when they may be neglected over whole range (as being $\leqslant 4$ units in last decimal).

3. $f(4 \cdot 2) = 0 \cdot 238095$. Throwback of fourth on to second differences is not good enough because of the large interval and number of decimal digits; use Everett to sixth differences.

4. $\delta^2 f$ is about 10, and makes a contribution to Bessel's formula of about one unit in the last decimal over much of the θ-range! Values for the tangents using Bessel's formula are 2·36158, 2·40827, 2·46888, and using mean differences are 2·36181, 2·40831, 2·46866, respectively, and thus errors of *more than twenty units in the last decimal* are possible near 0' and 60'! In fact the values using mean differences for 67° 3' and 67° 57' are correct to 3D only!!! and that for 67° 27' to 4D! It is hard to see why the author retains mean differences for 67° but rejects them for 68°; and the reader may find it instructive to investigate other short sets of tables and to compare them with Chambers's *Shorter Six-figure Mathematical Tables*.

5. First reaction is to use the Newton-Gregory forward difference interpolation formula as we are working near the head of a table, but we find that the terms decrease too slowly for effective calculation. (*N.B.* if we take many terms this will involve many tabular values, only a few of which will be from the neighbourhood of $x=0$, where $\tan \sqrt{x} \backsim \sqrt{x}$.) We can avoid this difficulty by *squaring* each term in the table and working with $\tan^2 \sqrt{x}$ (which behaves like x near $x=0$); after the interpolation has been carried out successfully for $\tan^2 (\cdot 075)$ take the square root.

6. Use divided differences for whole subtabulation, or alternatively to tabulate at $x = 0(\cdot 3)2 \cdot 4$ followed by Newton-Gregory at constant interval. Former method probably simpler once divided differences have been calculated. Check final table by differencing. Actually $f(x) = \cosh x$ and solution may be checked from tables, though it is essential that the computer should gain confidence in his own checks for ensuring accuracy.

7. Everett remainder same as Bessel; Stirling remainder term is

$$\frac{\theta(\theta^2 - 1^2)(\theta^2 - 2^2) \ldots (\theta^2 - n^2)}{(2n+1)!} h^{2n+1} f^{(2n+1)}(\xi).$$

ANSWERS AND NOTES ON EXERCISES

Chapter Four

1. Differentiate Bessel's interpolation formula; use result $f' = -f$ to check result by interpolation on table given.

2. Obtained from finite difference formula by breaking down the first, and the first two terms, respectively, into tabular values.

3. (ii) $-h^7 f^{(6)}(\xi)/140$; (iii) $-5h^3 f''(\xi)/6$ compared with $-h^3 f''(\xi)/12$ for each application of the trapezoidal rule and $-h^5 f^{(4)}(\xi)/90$ for each application of Simpson's rule. Dufton's rule is comparable with the trapezoidal rule in accuracy and is remarkably easy to apply: it is useful for a quick estimation of an integral from a table.

5. Correct value is 0·86638.

6. Correct value is 0·37558.

7. Obtained by integration of Stirling's interpolation formula. Second form for integration over a double interval obtained by adding formulae of first type for
$$\int_0^h \left(\int\right) + \int_{-h}^0 \left(\int\right) = \int_0^h \left(\int\right) - \int_0^{-h} \left(\int\right),$$
and last term obtained by suitable modification of original formula.

8. $\int_{-h}^h dx \int_{-h}^h f(x,y) dy \simeq \tfrac{1}{3} h^2 \{8 f_{00} + f_{11} + f_{-11} + f_{-1-1} + f_{1-1}\}.$

Matching is quadratic, and correction term is
$$-\frac{1}{45} h^6 \{f_{xxxx}(\xi, \eta) + 10 f_{xxyy}(\xi, \eta) + f_{yyyy}(\xi, \eta)\}.$$

Chapter Five

1. $16 - 1\cdot3x$. Note that the rms residual error is about 1, and there is nothing to be gained from specifying the line more accurately.

2. $5\cdot85 x^{332}$, with rms residual error of about 0·03.

Directory of Numerical Processes

Polynomial approximation

Exact matching of equidistant tabular values (see Interpolation)
Least squares matching of discrete tabular values: §§ 5.1, 5.2.
Continuous approximation by least squares: problem 6 (§ 5.2).

Other approximations

Exponential and power law approximation by least squares: § 5.2.

Interpolation

Using divided differences based on table with variable interval: § 3.2.
Linear interpolation: § 3.1.
Using differences of a constant interval table: §§ 3.4, 3.5; including throwback: § 3.5.

Differentiation

Using differences of a constant interval table: § 4.1.
Using tabular values only: exercise 2 (Ch. 4).
Half-way formula using finite differences: exercise 1 (Ch. 4).

Integration (Quadrature)

Using ordinates at constant interval without differences: § 4.2.
Using constant interval differences: § 4.3.
When the integrand has a singularity: problem 4 (§ 4.3).

DIRECTORY OF NUMERICAL PROCESSES

Double and repeated integration: § 4.4, exercise 7 (Ch. 4).
Using ordinates at 'optimum' points: § 4.5.
Using divided differences: exercise 4 (Ch. 4).

Other processes

Detection of errors in a table: example 2 (§ 2.5).
Extraction of roots of an algebraic equation: example 8 (§ 3.6).
Interpretation of a number: § 1.3.
Line of 'best fit' for (experimental) data: § 5.1.
Smoothing of irregular data: § 5.3.
Summation of series: problem 5 (§ 4.3).

Index

Accuracy of a number, 7

Backward differences, 21
Bessel coefficients, 42, 52
Bessel interpolation formula, 42, 44
  remainder, 55
Blunders, 10, 11

Central differences, 21
Continuous approximation, 90
Continuous functions, 8
Critical tables of coefficients, 52
Curve fitting, 9, 83

D, 15
Data smoothing, 90
Decimal digits, 7
δ, 22
Δ, 20
Detection of errors in tables, 25, 26
Differences, 17, 18
  backward, 21
  central, 21
  divided, 28, 29, 30
  expected fluctuations, 20
  forward, 20
  modified for throwback, 46, 51
  of a polynomial, 22, 23, 29
Differentiation, 56, 57
  numerical processes, 59, 80
  errors, 60
Divided differences, 28
Double integration, 75, 77
  Simpson's rule, 75
Dufton's rule, 81

$\eqsim$, 6, 37
Error fan, 25
Errors, 10
  detection, 25
  in numerical differentiation, 57, 60
  in numerical integration, 64
  probable, 10
  random, 9
  round-off, 10, 19
  systematic, 9
  truncation, 11
Euler-Maclaurin summation formula, 73
Everett interpolation formula, 42, 44
Exponential approximation, 88
Extrapolation (Newton-Gregory formula), 40

INDEX

Finite differences (*see* Differences)
Floating point decimal form, 8
Forward differences, 20

Gaussian quadrature, 78, 79

Integration, 58, 62
  divided difference formula, 81
  Dufton's rule, 81
  errors, 64, 66
  Gaussian, 78, 79
  Simpson's rule, 64, 68
  three-eighths rule, 64, 80
  trapezoidal rule, 63, 67
  Weddle's rule, 80
Interpolation, 8, 32, 33, 56
  Bessel formula, 42, 44
  choice of process, 44, 47
  divided differences, 34, 36
  Everett formula, 42, 44
  half-way formula, 43
  inverse, 49
  linear, 32
  Lagrange formula, 38
  Newton-Gregory formula, 39
  Stirling formula, 43
Inverse interpolation, 49
Iterative process, 51

Lagrange interpolation formula, 38
Least squares, 84
  continuous approximation, 90
  exponential approximation, 88
  linear approximation, 84, 87
  polynomial approximation, 85, 87
  power law approximation, 88
  r.m.s. residual, 85
Linear interpolation, 32

μ, 42

∇, 21
Newton-Gregory interpolation formula, 39
Newton interpolation formula, 36
Normal equations, 86
Notation for finite differences, 20
  for tables, 15
Numbers, 7

Operational methods, 93

Periodic functions, 16
Pivotal values, 48
Polynomial representation, 3-6, 24
Power law approximation, 88
Proportional parts, 34

Quadrature (*see* integration)

INDEX

Random error, 9
Round-off errors, 10, 19, 20
Rounding-off procedure, 10

S, 15
Semi-empirical theories, 53
Significant figures, 7
Simpson's rule, 64, 68, 75
Smoothing, 9, 90
Stirling interpolation formula, 43
Subtabulation, 48

Summation of series, 73
Systematic error, 9

Tables of values, 15
Tabular interval, 15, 20
Taylor series, 2, 53
Throwback, 45
Trapezoidal rule, 63, 67
Truncation error, 3

Weierstrass's theorem, 4